Shaped by Vegetal Matters

Critical Plant Studies

Series Editor

Douglas Vakoch

This series calls us to re-examine in fundamental ways our understanding of and engagement with plants, drawing on diverse disciplinary perspectives. The series encourages work grounded in the humanities and social sciences that provides innovative reformulations of the scope and practice of critical plant studies. Books in this series include both monographs and edited volumes that target academic audiences. To introduce critical plant studies to readers not familiar with this field, the series publishes work that is relevant to those engaged in critical plant studies, while also being of interest to scholars from the author's primary discipline. Among the books of special interest for the series are those that examine plants with reference to particular countries or regions of the world, or with respect to specific cultural, philosophical, religious, or literary traditions. Contemporary and historical works are equally appropriate. We especially welcome books that bridge academia and activism.

Recent Titles in the Series

Shaped by Vegetal Matters

Phyto-Influence on Humans, Other Animals, and Place

Elizabeth Oriel

LEXINGTON BOOKS
Lanham • Boulder • New York • London

Published by Lexington Books
An imprint of The Rowman & Littlefield Publishing Group, Inc.
4501 Forbes Boulevard, Suite 200, Lanham, Maryland 20706
www.rowman.com

86-90 Paul Street, London EC2A 4NE

British Library Cataloguing in Publication Information Available

Library of Congress Cataloging-in-Publication Data

Names: Oriel, Elizabeth, author.
Title: Shaped by vegetal matters: phyto-influence on humans, other animals,
    and place / Elizabeth Oriel.
Description: Lanham: Lexington Books, [2025] | Series: Critical plant studies |
    Includes bibliographical references.
Identifiers: LCCN 2024041647 (print) | LCCN 2024041648 (ebook) |
    ISBN 9781666940527 (cloth) | ISBN 9781666940534 (ebook)
Subjects: LCSH: Human-plant relationships.
Classification: LCC QK46.5.H85 O75 2025 (print) | LCC QK46.5.H85 (ebook) | DDC
    581.6/3–dc23/eng/20241029
LC record available at https://lccn.loc.gov/2024041647
LC ebook record available at https://lccn.loc.gov/2024041648

∞™ The paper used in this publication meets the minimum requirements of American National Standard for Information Sciences—Permanence of Paper for Printed Library Materials, ANSI/NISO Z39.48-1992.

To Sandra, Arthur, Marina, Daniel, Fiona, and Hershel.

Contents

Acknowledgments

I first acknowledge the plant kingdom. Plants structure the world, both physically and in numerous other ways that this volume explores. Humans are nested inside a vegetally structured world, such that plants are foundational to human lives, cultures, and experiences. I am grateful to the entire plant kingdom and am a passionate lover of vegetal beings.

The ideas presented in each chapter came about organically, without a prior plan. Each chapter had a somewhat independent origin, though the theme of being attentive to other species' interests and contributions has been a lifelong theme.

The central findings in this volume began to take shape during my PhD research at the University of London. Funded by a Bloomsbury Scholarship and aided by J. Simon Rofe and Ayona Silva Fletcher, my ethnographic fieldwork opened me to the dynamics of human, other animals such as elephant, and plant relations in Sri Lankan landscapes. When not on fieldwork, walking among London's ancient trees shaped understandings of plants and their roles.

The chapter on ash trees was developed with my co-author, Anna Perdibon. The willow chapter came about with generosity and learning from Ane Lyngsgaard. I am grateful to both.

More recently, I want to acknowledge the Aarhus University project called "Seasonal Journalism as Vernacular Phenology," funded by Aarhus University Research Foundation, and project leaders, Henrik Bodker and Michelle Bastian. The jacaranda chapter received support and input from this project, while in general, this volume owes much to the support of this collaboration.

Introduction

Shaped by Vegetal Matters: Phyto-Influences on Humans, Other Animals, and Place

Plants may be far more exquisite ciphers of "place" than the mammals examined by ethologists and ethnographers.

(John Hartigan 2019:2)

On a hot October day, I visit farmers on a plantation. We stand in fields amidst millions of sugarcane plants. I am in southern Sri Lanka—a small, teardrop-shaped island nation southeast of India, and it is 2019. As I move reddish brown earth with my sandals, the soil scatters like dust. Farmers tell me the soil is dead, as cane is grown year-round and agrichemical inputs are enormous. Sugarcane is a thirsty plant, and as such, this landscape of cane extending for 67,600 acres extracts heavily from groundwater, leaving less water for others. Perfume-like scents move through, a synergy of the cane and invasive plant species nearby, and I hear flies buzzing. The canes' grassy stems are pale green; with all the life nearby, the atmosphere does not feel vital. Thinking with plant life, as Hartigan (2019:2) suggests in the epigraph, is attending to secret languages of ecocultural histories and placemaking.

Just over the road is a national park where Asian elephants move and forage, though they prefer to eat sugarcane over diminishing grasses in the park. Certain males enter the plantation at night. Some of this cane will be processed into arrack, a potent alcohol that farmers (and even elephants at times) consume, seemingly to self-medicate. Many speak openly about their severe emotional stress caused by the local conditions. Teak trees, another thirsty plant, grow in rows in another plantation run by the government at the edge of the park. These adjacent spaces—national park, plantations, village, and small-scale farms—house different plant agencies, relationships, colonial and pre-colonial histories, and uses; each is distinct, unrelated to the others.

1

Plants have created habitability for animals across landscapes in contextual and interconnected ways. Yet these plantations, designed in differentiated, unrelated spaces, are not fostering habitability and local thriving. Farmers in this region, both on plantations and on small farms, struggle to make ends meet with wildlife eating their crops. Meanwhile, elephants suffer from reduced access to palatable vegetation. In contrast, traditional cultivation (vegetal) spaces were distinct but related, with areas of upland forest, vegetable crops, rice or paddy, water ponds, and spaces for humans and other species. Each area of vegetation had roles for diverse species' needs while supporting the whole system to prosper. I see how the current vegetal design fosters societal fragmentation and conflict amidst humans and elephants. Plants and human-plant relations, because they are so foundational to life, orient other relationships, such as human/elephant and human/other animals. In this way, plants are shapers of relationships and are beings of context, exerting contextual influences.

This volume traces lines of influence, affect, and alliance moving across vegetal toward animal lives and places, opening up multi-directional influences. Each being here is a *situated subject in a multispecies social milieu*. The boundary lines here are drawn more inclusively than the dominant ones today, which are drawn around human self, family, city, neighborhood, and/ or nation. Care for plants, their subjectivities, their needs and interests, and addressing the extinction and deforestation crises require new ecological politics and legal frameworks. Extending the social milieu in both perceptual and material ways, such as in policy frameworks, requires a sense of vegetal person-ness, agencies, influences, and significance.

Influence moving between plants and humans is generally recognized in how humans influence plants in contexts such as agriculture and food systems, forestry, and deforestation. As important as each of these spheres is in expressing human-plant relations, and with each being central to the current climate crisis, they are only half of the picture. The other side, namely vegetal influence, is obscured and hard to perceive, let alone discuss. The success of modernity in perpetuating the perceptual gulf between humans and nature operates through obfuscation and denial. New approaches to plant relations that shed light on vegetal roles and capacities open space for new ways of engaging with ecological degradation.

This sensibility to plants as agents, as subjects, is apparent for many "ecosystem people," a term from authors Madhav Gadgil and Ramachandra Guha (2013) that describes those who live from their land base.[1] During my fieldwork in Sri Lanka, most people I met were able to name and speak of the medicinal qualities of numerous wild-growing plants. Those living in cities generally possess the least intimate plant sense and knowledge (Muratet 2015), and yet urban spaces are generally where land use decisions are made for those in rural areas. In terms of loss of plant knowledge and a shift in orientation to what has value, American teenagers can identify hundreds of corporate logos and yet cannot identify ten plants (Hawken 1993). As Bruno Latour states in an interview, we have long "rendered earthly conditions invisible" (Pedersen et al. 2019:217).

Each case study in this volume seeks to redress this loss of attentiveness, appreciation for, and relationship with plants. This loss, which is evident in the phenomenon of plant blindness, is not only about the loss of ability to see and name plants, but also the loss of awareness of their influences. It is as if industrialized humans and their nervous systems have been trained in their sensitivities and sensibilities to be unaware of earthly and vegetal influences. And yet, across the chapters, five case studies of plant influences and an inclusive social milieu emerge from different regions of the world. From wapato on a river island in the United States, to willow in Denmark, to sugarcane plantations in Sri Lanka, to ash and jacaranda trees, features emerge of human-plant social intimacies, power dynamics, and intersubjectivities. Vegetal influence is accessed in diverse styles, analytically and descriptively, reflecting the vegetal research projects each emerged from and their diverse approaches.

Amidst highly domesticated cultures of industrial and urban worlds, recognition of vegetal influence and agency is building. Critical Plant Studies, the relational turn in social science, and pioneering scientists such as Monica Gagliano and Suzanne Simard have enlarged recognition of vegetal intelligence and complexity. John Hartigan's (2017) relational attempts in "How to Interview a Plant" and his project, Social Theory for Non-Humans, open up vegetal personhood. Limitations of our modes of study emerge. The social worlds of the vegetal appear with plant geneticist Barbara McClintock's work, who spoke of knowing a plant intimately, having a "feeling for the organism." Hearing grass scream when injured, which among other findings, led to rebukes from the scientific community. Over McClintock's career, she changed her focus from plant genes dictating their environment to the cell matrix as dominant. This cell matrix approach speaks to the centrality of context, of social milieu in dictating one's life. This volume takes up plants and their relations with context as a central force in vegetal influence.

To reiterate, a primary theme across the chapters is how *plants are in social relationships with other species, including humans and as such have affective roles and influence on human subjectivity and collectives. Further, plants are beings of context and as such influence and inform humans in their relations to context, to the interdependencies of place, and to interspecies coexistence and conflict. Each chapter also attends to making and poetics (poiesis) as a glue of phyto-human social worlds.* In each chapter, plants exert relational influences on land and co-design place, and, as such, are significant organizers of space and land into sites of culture, meaning, and patterns of coexistence for humans and other animals.

The research presented here worked with both "planthropology," as Natasha Myers (2017) recommends, in which plants are subjects along with humans and co-creators of culture, and with more-than-human ethnography as a method (Barua 2014). One chapter uses textual analysis of newspaper articles. Plants are excellent guides in a place-based, relational, diverse social milieu. If the earthly wastelands of human exceptionalism are actually to be renewed, and other beings recognized for their contributions to living systems as well as having sentience and intention, attention and listening to other species and their qualities and influences is a core practice.

The central theme of this book grew out of my own close attention to oak, service, and hop hornbeam trees, and to plantain, dock, and mugwort and their milieus, growing in the urban forests of London. Much like a psycho-geographer in London, who moved through cities, feeling drawn to places of heightened experience, I was drawn to certain trees and plants and looked forward to daily visits. I began to feel plants possessing qualities that moved out into the world, creating atmospheres and shaping the feeling of each place. I met very few others who observed trees closely; unless plants have meaning, either symbolically, aesthetically, or usefully, they are not regarded as worthy of attention, John Hartigan (2017) writes. These visits also left me grief-stricken, seeing the effects of global heating on my favorite oak trees while remembering a study showing how they are adapting through resistance to embolism in their xylem (Skelton et al. 2021). As such, my experience revealed plants as place makers with authority on living through crisis.

Walking on London's Hampstead Heath one day with an Indigenous leader from the Mata Atlantica region of Brazil, who happened to be staying in the same flat, opened more vegetal vistas for me. He stopped frequently to sing to the trees, maintaining and engendering relations with the vegetal, and with the wider world. I could see that these songs were part of a code of conduct. A robin's sing-song lyrical notes and a crow's raucous call joined his melody, creating a rhythmic syncopation. Trees seemed to perk up, their attention riveted to his song. Another leader from the same tribe explained that plants are more than persons, they are masters. He described his singing as generating the world into being, maintaining the world. His singing to plants can be seen as interspecies sociality and culture.

Singing poetry to plants had been central to small-scale, subsistence farmers' practices in Sri Lanka, even up to 50 years ago. "We have a poem culture," one agronomist told me, explaining how poetry sung to plants and to animals maintained good relations. Poems pleaded with elephants to eat crops, and warned plants and animals about possible harms caused to them when farmers burned the fields. Thus, poetic songs were part of a trans-species social and cultural milieu that has been lost in many places to technology, agrichemicals, mechanized practices, and terraformed land.

Plant cultivation involves poiesis as a way of bringing food into the world that, in the Sri Lankan context, was aided by the use of poetry. Poiesis is an ancient Greek term, referring to making something new in the world, as a craftsman does; the word "poetry" also comes from the same root. The poems sung to plants are also a poiesis, enacting social codes with others in the world. Poiesis is both active and passive, involving skill and sensitivity to one's material and surroundings, and not only produces objects of craft but also produces meanings, which can be thought of as forms. Dreyfus and Kelly's (2011) *All Things Shining: Reading the Western Classics to Find Meaning in a Secular Age* suggests that poiesis is a kind of antidote to what has been lost in secular and technological society.

In the "Poetics of Cohabitation," Ribo (2022) describes how oral poiesis in the form of song is central to cohabitation across species and has been central to many Indigenous practices on the land. Poiesis in this volume straddles the ledge of skill and poetry and refers to them both as modes of relating to other species. Across the case studies in this volume, human-plant social worlds are mediated by language and works of poiesis.

In a mysterious way, this book seemed to spring or unfurl to life in an organic way that suggests an interspecies poiesis. Opportunities opened up unexpectedly, allowing new chapters to come to life. Plants seemed to have a role in propelling the process along, though to be more precise, the process evolved in the in-between spaces. For instance, living in a basket-making community in Denmark was quite remarkable and positioned me to study willow as an agent of and in basket weaving.

A sense of plants as place makers also grew through observations of a small plant, yerba mansa *(Anemopsis californica),* growing along the Rio Grande River in New Mexico. While walking amidst cottonwood trees and so-called invasive Siberian elms and Russian olives at the river's edge, when you reach the large colony of this perennial, yerba mansa's camphoraceous scent envelopes you. The volatile oils are reminiscent of thyme, basil, or eucalyptus. They improve soils, reduce salinity which is increasing through deforestation and global heating, and they remove toxins, while also being a potent medicine for humans. This plant has a presence, as does another potent medicine, snowdrop (*Galanthus* species), a similarly small but powerful flowering plant that breaks through snow cover across Europe in late winter and is used medicinally to treat Alzheimer's disease. In their improvements to ecological conditions and in their medicines, they exert a personality that expresses itself through their qualities. Yerba mansa's aroma on the river is an aspect of the plant's personality. Experiential moments in the plant's presence called on me to look for patterns across species that can be learned from, that foster continuity and life.

SEEING PATTERNS

Observing interspecies lines of phyto-human alliance and influence grew and took shape not only in London parks or along the Rio Grande River, but also from my 2018–2019 ethnographic fieldwork in Sri Lanka, studying human-elephant conflict. Asian elephants *(Elephas maximus maximus)* on this tropical island have for millennia foraged on crops, yet in the last decades are dependent on crops while coexistence with farmers breaks down, with many more deaths. In 2023, 470 elephants died, of which half came from conflicts with humans, while 176 humans died from the conflict (from Dept of Wildlife Conservation data). An accurate count of elephants is challenging, but between 5,000 and 6,000 likely inhabit on the island. Living with elephants on a small island highlights socio-ecological issues, as space is limited and tensions mount quickly.

The Great Acceleration, which refers to the last approximately 70 years of expanded extraction and production, has altered the workings and logics of human-elephant-plant relations in Sri Lanka. Areas of cultivation expand, taking over remnants of forest cover, and monocrops in plantations have replaced situated relations in traditional slash-and-burn cultivation called *chena*. Eight large dam and reservoir projects have redesigned large swaths of land. Elephants now move between small forest patches in some regions, foraging on garbage heaps; some say elephants are now homeless and tame, having lost their terrain which gave them autonomy.

My research traced changing relationships between humans and elephants over a 70-year period in two sites in Sri Lanka, though more distant histories were also considered. Yet, fairly quickly, I came to see plants as situated right between these two animals, humans and pachyderms, providing a mediating force in how these two coexist. The ways that humans relate to plants in Sri Lanka, which is covered in chapter 1 of this volume, extend to and shape relations with other animals. This discovery, which highlights the lack of distinctions between kingdoms and between beings, and the relational nature of the nature/culture spheres, makes biospheric and ecological sense.

Thinking on a systems level, plants are ecological *producers*, creating food and habitability across the globe. Vegetal life fosters diversity in 383,054 known plant species (worldfloraonline.org) and fosters abundance as they make up 80% of Earth's biomass. Plants generate 98% of the oxygen that humans breathe. In contrast, humans and other animals are ecological *consumers*. This consumer-to-producer relationship is a primary one, and plants are axial agents in how humans exist on land and relate to their context. Plants in the form of crops in agriculture now cover 30–40% of the Earth's surface, account for 98% of fresh water usage (Huang 2019), and form the largest ecosystem on the planet. Meanwhile, irrigation for agriculture *uses* 70% of the

Earth's groundwater (data from Project Drawdown). This terraforming with plants alters plants' relation to context, as so-called invasive species dominate terrain, with altered modes of diverse generativity.

Situating the reader with me, on the ground in Sri Lanka, I watch as a female elephant herd moves slower even than a leisurely pace through a national park at 7 a.m. on a sweltering day in the south. Their grayish bodies have been rolling in mud to protect against insects and heat, and so they now mirror the soil beneath them in rich reds and brown earth tones. Scents of damp mud are held in the wet air with our own breath, and I notice my own clothes are drenched as usual in sweat. Grabbing very young teak trees which have taken root here, these pachyderms seem to relish the large juicy leaves and avoid most other small plants. Teak trees are a high-priced commodity here and a very thirsty tree, drawing heavily on groundwater, as does sugarcane. In the nearby teak plantation, elephants pick away at the bark when hungry.

A profoundly sweet scent overtakes others, emanating from a bright pink and orange-blossomed invasive plant, *Lantana camara*, or butterfly bush. Introduced to Calcutta, India in 1809, this ornamental now grows everywhere, exiling native species. Between the scents of Lantana and sugarcane from a nearby plantation, I wonder if elephants have trouble orienting themselves to place, as they possess robust olfactory abilities and use scent to navigate. The monocultures of plantations that began with colonial rule on this island are concurrent with an olfactory monoculture, with sweet scents of monocrops and invasive plants overtaking other, more subtle ones.

This region can be mapped according to plant scents or to elephants' vegetative preferences. Humans have certainly mapped these lands in various ways, with colonial maps drawing hard lines and boundaries of property lines that had not existed prior to foreign interference. And yet, a different kind of mapping is possible, one that attends to plant relations with humans and elephants and how these foster coexistence or conflict. Human-plant-elephant relations engender one another, with plants setting the stage, the vegetal matrix, and thus being foundations to all animal relationships.

In this fraught and conflicted context, I came to see the centrality of plants and their socio-ecological roles. Plants were the central actors in cultivation and in elephants' foraging patterns, as well as in how water is held by the dry soils, generating climate and microclimate. Human relations with plants in cultivation, in forestry, in hydrology, in identities, and in worldviews set the tone for human-elephant relationships.

Thus, I came to witness and participate somewhat in a human-elephant-plant social milieu. This milieu has certain codes of behavior and even culture, in how practices build rapport. Over millennia, elephants have accessed crop remnants after harvest. This millennia-old agreement between farmers and elephants is a trans-species social contract and shapes a politics of coexistence. Humans and elephants shared terrain, though with different periods of access (Fernando 2000; Oriel and Frohoff 2020). Across time and space, rhythmic periods and spaces of access and no access were accepted to some extent. Songs and poems were sung to protect crops and also to warn animals and ask permission or forgiveness of plants before burning fields in slash-and-burn cultivation. All of this was part of a trans-species culture and poiesis. These dynamics were very localized arrangements of power and distribution, in which farmers have had some autonomy over how they work.

In the last 70 years, that autonomy has broken down, as has the poem culture and human-elephant coexistence, with conflict being the norm. Coexistence is a choreography of *social relations across species and place*, a sociality that is completely denied to other species in dominant modernist systems. In the Sri Lankan context, these interspecies social relations are apparent, involving experiences, practical knowledge, learning, emotions, and a sense of other species as sentient and intentional. Practical knowledge is built by *learning from place*. This approach is starting to take hold in the relational turn in social science that views each being as a subject with agency emerging from in-between places.

PLANT SUBJECTIVITIES

It would be difficult to overstate the significance of plant life to other beings. Humans exist because of forests. Rivers exist because of forests. They are the foundation of animal life and direct the planet's water cycle. And yet current rates of deforestation are not in sync with this basic understanding. Furthermore, forests are not just functional for animals; they are also cultural and tied to meanings inscribed in place and time. The future of animal life rests, in large part, on relations with plant life. Plants, in their situated and co-evolved contexts, have been the apotheosis of diverse generativity, embodying a creative principle. Yet they also foster complexity in their modes of relating with humans, in food, medicine, clothing, homes, household items, as spiritual guides, ancestors, in their watershed and riparian roles, and in roles of creating food and habitat for many species. Vegetal presence is so fundamental to human and other animal lives that exploring their modes, relationships, and ways they influence humans is a fundamental exercise in future-making.

A phyto-human social world is one that Myers calls up in "planthropology," which recognizes plants as subjects. Being a person or a subject in the Western tradition relies on a certain measure of intelligence. Recent research in plant cognition by Anthony Trewavas, Paco Calvo, and Monica Gagliano, among others, reveals plants' abilities to learn, remember, communicate, and collaborate. Yet other cultures outside the European Enlightenment tradition (or ones that informed the Enlightenment[2]) hold to more-than-human subjectivity in different approaches. For example, plant subjectivity is expressed by one interviewee in chapter 5 about ash trees. Tanhagha Yako, an Anishinaabe tradition holder, explains that in her culture, there are more than four directions–there are actually seven. The seventh is "the fire within, where we learn how to move from love and compassion." Trees, she said, have this seventh direction also. This shared sense of internality and subjectivity is recognized less through IQ measurements and more through abilities to love and to treat others well. A concept of personhood as emerging not from rational intelligence but from emotion and ethics is echoed in British philosopher Mary Midgley's (2004) writings.

Another approach is with John Hartigan's (2019:1) question, can plants actually be considered ethnographic subjects. He writes, a "plethora of studies reveal that they [plants] are intelligent, agential, and social" (2019:1). Interpreting place-based dynamics across beings that are often central to ethnography can be remiss without including plants as subjects. And yet plants are complex to understand and access in their behaviors with such different modes of communication, such as their roots signaling through mycorrhizal networks and with phytochemical signaling across trees. One way into vegetal lives is in chemo-sociality (Shapiro and Kirksey 2017) that emerged as a branch of the relational, more-than-human social science lens, in which plants as subjects participate through their volatile oils. Though industrial humans have lost an attentiveness to plant presence, let alone scents and forms, their agencies are finding recognition in the burgeoning worlds of Critical Plant Studies.

An aspect of recognizing plants as subjects is to recognize their influence on the world, which is an expression of subjectivity. Plants have influences, the renowned Prussian scientist Alexander von Humboldt (2016) declared, through their appearances. In his exhaustive exploration of plants and their contexts in Latin and North America and Europe, he wrote how they exert emotional effects on those in their presence, while also expressing their relations with global processes of weather, topography, and aspects of processes inside the Earth. Inspired by Humboldt's trans-disciplinary work, and his argument that single disciplines cannot access nature's complexities, each case study situates plants in complex relations in a living meshwork[3] across beings, land, climate, meaning making, and culture. A vast interdependence renders nature a unity with vast diversity, Humboldt argued. This paradox of unity and diversity is a theme across plant lives and influences.

PLANTS STORIES AND KNOWLEDGES FRAME
RELATIONALITIES

Plant influences on humans and other animals work on the physical level, the subjective and the community level, with plant stories as a form of phyto-human culture or poiesis, carrying central logics and relational styles. As Felix Guattari (2005) writes in *The Three Ecologies*, the social, subjective, and ecological are not separate spheres but are mirrored and indistinct. For example, the presence of a forest providing for animals' diverse needs is a physical quality, and the same forest carries meanings, practices, and rituals that speak to subjective and socio-cultural qualities.

Plant stories in most cultures are central to making sense of the world, and thus frame social worlds across beings. For example, the ash (Fraxinus species) is the World Tree in Celtic and Norse stories, holding all the worlds together. The Bo tree is a sacred plant in Sri Lanka, as Buddha attained enlightenment while sitting underneath one. Their huge trunks and graceful bodies grace Buddhist temple grounds on the island, and many worshipers present them offerings. These examples of phyto-human cultures involve stories, practices, ontologies, and histories that are in dialogue. Theoria, praxis, and poiesis all in dialogue.

An example of plant actors in human origin stories that reveal and frame truths can be found in Genesis in the Hebrew Bible. The two tree actors in the Garden of Eden offer up an analogy for present times. Relations with these two trees dictate the first humans' ability to remain or be expelled from the Garden. God told Adam and Eve that they could eat from the Tree of Life, which brings immortality, but are forbidden to eat from the Tree of Knowledge of Good and Evil, which leads to death or a loss of immortality and expulsion from the Garden. One can interpret the Tree of Knowledge of Good and Evil as representing a dualistic approach to the world—one that divides good from evil, humans from nature, bodies from minds, and nature from cultures. This worldview of separate and distinct realms has allowed for industrialized worlds and ecological destruction, as nature is positioned separate and outside of human life and is thus less valuable or without inherent value. This divide is also apparent in fragmented and distinct approaches to land use and spaces, as mentioned in Sri Lanka.

Fragmentation also takes place in siloed pillars of plant knowledge and in human-plant relationships in general. Here is an example of divergent realms of nature and culture, in which science dominates as a way to know plants, while story is in a separate realm of art and culture. I joined a local synagogue's celebration of a Jewish holiday, Tu Bishvat, a new year for the trees, in a walk on Hampstead Heath in London, led by an artist, rabbi, and ecologist. The rabbi quoted from Exodus on how Moses parted the Red Sea and put a plant in the water to sweeten it. The ecologist spoke of trees in the language of statistics and measurement. The artist read a few poems. Though the event brought these knowledges together, it was also apparent that these languages are distinct and do not speak to each other or interact.

The scientific holds the most credibility as an access point to vegetal lives. Yet, as Amitav Ghosh (2021) says, "plants are usually only viewed from one side which is science and commerce, yet they have a whole other half that is known in story and song." This helps explain the loss of attention to plants and plant knowledge, for they hold little interest through the dry facts of botany.

Ghosh argues for the arts to reimagine and recognize more-than-human subjects. The phenomenon of plant blindness (Wandersee and Schussler 1999) (where, in experiments, people fail to see plants in favor of animals, such that plants are either viewed as the backdrop or are invisible) can be viewed as a broken relationship between industrial humans and the vegetal. An aspect of the breakage is the distinct knowledges that are not in dialogue.

As industrial societies choose the pursuit and logics of dualistic thinking, good and evil, instead of the logics of the living world, the interconnected crises of climate and extinction persist and accelerate, and death is nigh. The living world, as expressed by the Tree of Life, works through logics that are relational, reciprocal, and interdependent, with each engaged in supporting the system. The Tree of Life supports the immortality of life, as all the beings on Earth regulate the atmosphere and temperature through each being's life cycle, for the continuity of life. James Lovelock and Lynn Margulis revealed this incredible interdependent web in the Gaia Hypothesis in the 1970s. The choice between two Garden trees persists, yet the time frame to change course diminishes rapidly. Learning from plants and their situated relationships means engaging with the logics of the Tree of Life. Poiesis creates new stories, new meanings, new objects, and through phyto-human skills, provides throughways to different knowledges and approaches.

VEGETAL AGENCY, PLACE, AND CONTEXT

Beyond the realm of plant stories and subjective framings, plants influence animals on the community or societal level through the ways that trees and plants exert their contextual agencies and co-construct place. Plant agency is situated in contextual relationships. Agency is challenging to define, let alone determine which species have this capacity. On a basic level, agency is the capacity to act, and yet philosophers, psychologists, human-animal scholars, and others debate whether it involves morality, intentionality, awareness, and consciousness. Echoing through certain scholarship within botany (Gagliano

2015) and science and technology studies (Latour 2005) is the notion that agency resides amidst agents, in the relationships between beings.

Gagliano cites the work of Chilean biologist Humberto Maturana, for whom cognition is situated within ecological contexts and resides in the activities and behaviors within systems of perception, adaptation, decision-making, and more. If plants have agency, their work is subtle, sensitive, and operating on different time scales than human beings. For this reason, Fleming (2017) suggests attending to "plantiness" or to plants' capacities and ways of being. This project adopts Karen Barad's (2007) agential realism, in which beings and phenomena do not pre-exist their relationships but emerge from them. One may certainly view plant agency in this light, as plants act in relation to conditions and to other beings in their environment, communicating through mycorrhiza, warding off insects by releasing chemicals, and sharing nutrients.

Taking a deeper probe into more-than-human agencies and capacities is anthropologist Anna Tsing's (2015) work on political, cultural, economic, and biological interactions of humans with matsutake mushrooms and Michael Marder's (2013) *Plant-Thinking, A Vegetal Life*. The matsutake mushroom is not only prized in Japanese culture, but their fungal capacities are extraordinary in their ability to nurture trees in degraded forests and aid in building them back. They work in the margins of capitalist destructions and offer hope for multispecies collaborative survival, Tsing suggests. Marder's philosophical work asserts that plants are different in how they are simultaneously locked in themselves and dependent on external factors, and yet also merge with the outside world, holding no delusional distinction between self and other, as humans do. His concepts draw on Levinas and Derrida in their critique of possessive selfhood, characterizing plants as respecting alterity and allowing others to pass through them without negating their otherness. He says they are teachers in the sense that they don't restrict themselves into stable identities.

Plant influences can be viewed as structured into their spatial organizations or geographies. Darwin argued that plant geography has been a keystone in the laws of creation. In other words, plants have organized the world for life to flourish, and embody laws of the living world. Yet plants, when removed from their situated, contextual relations, degrade diversity as seen in invasive species-dominated landscapes. For plants to hold such a central place in creation and to the principles of life suggests their influence on place, mind, and relations-in-place.

Plants, as central actors in and of the living Earth, are part of the big blur, part of what is obscured from view, from relationship at this moment, as modernity reaches a crescendo. While modernity and capitalism obscure, it is the work of individuals as activists to clarify and nurture vital phyto-human relationships, as a form of activism. Plants will out, as I discovered through a long winter living in London. With most of my human companions elsewhere, and on long walks in Big Wood, Queens Wood, Highgate Wood, and Hampstead Heath, I noticed trees and plants as central to my experiences, as shaping my experiences. This awareness had to do with an open space in my psyche, and with plants that moved forward to reveal their abilities and ways. This could be described as phytophenomenology, as Casey and Marder describe human-plant relations in *Plants and Place: A Phenomenology of the Vegetal* (2023).

Places are shaped by relationships, and plants are primary actors in place-making, as I found in London's parks, some of which are inhabited by ancient trees. As Marder and Casey reveal also in walks with plants, in their sessile though non-passive ways, plants live through immense attunement to the attributes and qualities of their relations-in-place and act accordingly, in where they seed, how they grow, and thus are contextual place-making. Casey and Marder bring plants' experience to the fore, through thinking with and through edges, and in how plants defy measurement and spatial coordinates.

This phyto-human social field is a contextual, situated field or place, thriving with soil, climate, other plants, insects, birds and other animals, and diverse microorganisms. This situated world is a phyto-social world (Casey and Marder 2023). Plants co-evolve with species in their environment, and each keeps others in check, limiting their expansion. Over long time periods, this process maintains diversity. These relations make up place, as a situated field of co-evolutionary relationships. The concept of place carries this relationality. Hugh Raffles poses an orientation to place as "continually and actively brought into being through the coming together of many human and non-human phenomena—physical labor, narrative, imagination, memory, political economy, the agentive biophysicality of tides, plants and animals . . . " (2003:px). These orientations undo the static, constrained human-centeredness with mind, imagination, and relationships moving across and generating each. The movements across place and beings in affect and influence are unique and singular to each species and place, offering guidance in ways the geo-social[4] proceeds.

As described earlier, plant life is so sensitive to context that when removed from its ecological site, it can mimic industrial ways of simplifying landscapes, as in Icelandic plains with non-native lupines (*Lupinus Nootkatensis*) from Alaska, or Sri Lankan jungles dominated by South American ornamental butterfly bushes (*Lantana camara*). In this way, plants expose the boldness of interdependence and relationality that underlies all life by extending human relational patterns of beings without contexts. The invasive examples above highlight plants mirroring human treatment of land and plant life, especially when vegetal life loses co-evolutionary partnerships. As such, the vegetal mirrors the human. These aspects are especially relevant in this time of genetically modified crops. Such modifications often prevent plants from making seeds, which means they are no longer reproducing and no longer evolving. Just as genes and evolution speak to context, plants are losing their abilities to connect to and speak of context.

As stated above as a central argument, plants are beings of context and thus exert contextual influences on humans. Due to their relative immobility, plants may perceive more environmental signals with greater sensitivity and discrimination than the roaming animal (Trewavas 2009:pxxx). Context is foundational to plant intelligence, which is apparent in their adaptability to the changing environment as an emergent quality across a system made up of a plant body within the plant's context (Calvo et al. 2020). These intelligences are not found in laboratory conditions, thus making controlled conditions less conducive to understanding plants' abilities.

Yet, places and plant communities are in flux at this time with climate and extinction crises, as many tree species native to myriad locales die, and with novel ecosystems forming. One example is in riparian or riverine areas along the Rio Grande river in New Mexico in the United States. Damming the river, and thus reducing the floodplain, is a kind of death sentence to native trees and plants that rely on pulses of water moving over land. With the added stress of drought and climate change, the matriarchal cottonwoods (*Populus deltoides wislizenii*) are losing out to the bold-looking, non-native tamarisk or salt cedar (*Tamarix species*), tree of heaven (*Ailanthus altissima*), and Russian olive (*Elaeagnus angustifolia L*). These new formations may be more attuned to future conditions, and yet they seem to be undoing the past; again, plants mirror human contexts and patterns, as modernity is itself a force that erases histories in movement toward an arrow (error) of progress.

The current ecological meta-crises can be viewed as essentially resulting from individual and collective *loss of orientation to context* in both time and space. The primary symbol of modernity in the arrow of progress has left the past behind, though the arrow swings wildly with no real direction, as progress shows itself to be a form of socio-ecological self-destruction. Meanwhile, contextual relations with place that foster continuity across beings have been eroded by capitalism's indiscriminate march toward capital gain and profit. Context, Karen Barad (2007) elucidates, is where relationships exist, in the spaces between beings. And thus, plants have roles in reorienting toward place and time.

SOCIALITIES, POLITICS, AND THEIR OBSCURING

In the academy and the media, plants are now described as having their own social life. "Plants have a social life too," says *Wired* magazine, citing how impatiens grow less root matter when surrounded by relatives as they share resources across kin in such contexts. The phenomenon of mother trees or hub trees supporting other trees and beings in a forest, described by scientist Suzanne Simard (2021), is an example of social relations in forests that include supporting seedlings through making space and symbiotic relations with fungi in mycorrhizal networks.

Anthropologist Anna Tsing argues for wider recognition of more-than-human social frameworks. The social, she writes, is "made with entangling relations with significant others" (2013:27), chiding the absurdity of thinking of other species as not social with each other. "Social relations are the forms through which the ways of life are organized" (2013:31). In her book, *A Mushroom at the End of the World* (2015), she calls relations between mushroom foragers, matsutake mushrooms, and pine trees a multispecies assemblage, with each cultivating the others. "Assemblage," a term from Deleuze and Guattari, describes how agencies emerge from human and non-human groupings.

Studies of complex human-plant relations in spatial relations or more-than-human geography help to blur the lines of nature-culture divisions and bring plant actors to the fore (Head and Atchison 2009; Barua 2014). Natasha Myers' (2020) art/research in a park in her hometown in Canada, attending to black oak trees and other beings, builds a sense of phyto-human socialities and how plants shape place.

In her work on companion species relations, Donna Haraway (2003) contoured a legacy for others to speak of the social, the ethical, and political across species. Her personal experiences help to explain what many know, who live with canine and feline companions—a hybrid sense of self across species, the myriad non-verbal, bodily communications, and sharing of beds and saliva. How can scientists hypocritically separate humans as social and the rest as not, while living in close relationships in which both are relational and intelligent agents? In some of the best work on place and extended social worlds, subjectivity guides in learning more than objectivity. These social worlds reside in subjective relationships, not in facts and objective truths. And yet knowledge about plant behaviors established through scientific methods does bolster one's ability to speak of these socially subjective worlds.

This volume takes up phyto-human assemblages as affective, social in nature, working at times through poiesis. This phyto-human social realm speaks to Natasha Myers' (2017:299–300) call for a "planthroposcene," in which humans recognize their "interimplication" with plants, and partner or make allies with plants and "change the terms of the encounter." "Plan-thropology" (2017) documents affective ecologies as key to human-plant relations, Myers writes. Case studies in this book speak directly to this call and vision.

Regarding a more-than-human sociality, Bruno Latour (2018) speaks of the "geo-social" as an earthly multispecies milieu, which requires a more-than-human, terrestrial politics and diplomacy. A wider benefit of a more inclusive social field is that ethics, justice, and politics apply across species and place in ways that the anthropocentric worldview of industrial life does not allow for. For example, until the seventeenth century in Ireland, the Brehon Laws protected trees with fines for harming or killing that were equivalent to fines for harm to humans. This legal system, rooted in an animist worldview, included trees as members of a social web. An example of shifting perceptions is apparent in perceptions of house plants, which became popular in the 1800s. In the beginning, they were considered living beings, but by the second half of that century, they were no longer viewed as alive, becoming decorative objects (Ruppel 2020).

In my fieldwork in Sri Lanka, studying human-elephant conflict, I observed and found examples of situated phyto-inspired socialities. One case in point is how grass inspires annual social gatherings for Asian elephants on the island. In groups of 300 or more during the dry season, when fresh grass emerges due to diminishing water levels in an ancient water body, elephants amass to eat their favorite forage and have a bold reunion with greetings, play, and frolicking in Minneriya National Park for 2–3 months. The relationship between elephants and grass is mirrored in farmers and rice plants (also a grass) in Sri Lanka, with a profound kinship connection (Van Daele 2008).

After all, for most of human history, trees have played a central role in physical existence in terms of food, shelter, shade, clothing, and also in experiential terms and meaning-making. The tree of life that inhabits myriad cultures' origin story connects all life, is the source of all life, and maps the various worlds in its branches. This intimate, kinship relation as foundational can be seen in present-day research on healing within green spaces. Leaving the hospital a day early after surgery and needing less pain medication correlates with a tree in view outside one's hospital window (Ulrich 1984). This vegetal influence is a direct one, measurable in medical norms. And one could argue that the view of the tree offers a relationship of context or even contextual lively and living support. Other studies on human relations with trees through interviews describe a sense of belonging, love, and connection to place in relation to trees in one's yard, along with other concerns like danger (Head and Muir 2005). Relations of humans with trees are anything if not confused at the moment, with the realm of nature considered to be "out there" and "other."

The stage is set for the recognition of a phyto-human social milieu as authors draw relational lines between plants and forests, place, knowledge, humans, and politics. Politics can be viewed as a realm made up of earthly relationships. In Sri Lanka, where I did fieldwork, forests are central players in politics, such that in their absence, human-wildlife conflict becomes a force of decimation to farmers' livelihoods and elephant well-being. This complex meshwork of spheres that we normally conceive as separate is interdependent in an inclusive social sphere. A linguistic example of plants as ontologically central to humans is found in the word "tree," which comes from a Proto-Indo-European language as *deru*, which meant true, and also solid and steadfast (Watkins 2020).

Plants' influences on animals are foundational yet obscured in the dominant value system of economic growth, in which social (eco-social) wellbeing is not prioritized and is sidelined. This sidelining is part of modernity, treating the living world as if it were not actually real. Life today in this heightened machine age can have a distinct flavor of the surreal, as the system seeks to convince that other vegetal and animal beings are not as real as what is inside a screen. Modernity fosters a perspective of oneself as distinct and separate from one's context. Yet each of us lives from a land base, though today this means living from numerous distant land bases, as tomatoes in our salads are grown on the other side of the world. These relationships are abstracted and hard to appreciate or understand (Head and Atchison 2009).

Keeping consumers from viewing the ecological wastelands left by industrial agriculture that perpetuate poverty in many regions is a primary vehicle for capitalist expansion. Social scientists researching global supply chains seek to undo this obscuring, revealing the power dynamics and injustices in globalized chains of interactions that bring shrimp literally across the world to a local grocery store in Wisconsin. Most urban dwellers have no connection to who grows their food or where their water comes from. In Sri Lanka, where I did my fieldwork, farmers grew food for their families without pesticides and yet, used enormous quantities on crops they sold. When supply chains are visible, as in family garden plots, ethics are in play, while the market exists with little earth-based ethics or regard for well-being.

One could argue that plants have their own "political" structures in forests, built around communication, protection, competition, kinship, alliances, and decision-making. Communication is a crucial trait in political relations, in community structure, and protection (Trewavas 2016). The communication among trees in a forest flows through mycorrhiza, and this flow enables trees to defend against predators, disease, and more (Gorzelak et al. 2015). Trees in a forest alert other trees to danger and ward off threats in a communal fashion. Marx's words, "society does not consist of individuals, but expresses the sum of interrelations, the relations within which these individuals stand"

(Marx 2005:265), could be equally true for ecosystems and forests. Mycorrhizal function and abilities are an interactive meshwork among individuals,
echoing Marx's words.

In this new more equalized and leveled field in which plants exert influence
and a kind of relational agency, how do human-plant politics play out in forests and in agricultural fields? Marx does have some thoughts on these questions. Though Marx's theories are criticized for ignoring ecological factors
and he certainly never mentions plants as political subjects, he does say that
agriculture is a product and instrument of capitalist modes, impacting both
farmer and soil. "All progress in capitalist agriculture is a progress not only
in the art of robbing the worker, but of robbing the soil" (Marx 1976:637).
Alienation is a state induced by the forces of capitalism, and Marx describes
the source of this as an alienation from nature. Describing urbanization's
impact on agriculture, Marx describes capitalism's "metabolic rift" as vegetables and fruits are transported to cities where waste products end up not
replenishing soil but leaving a gap, or rift, in the cycles of growth, decay, and
regeneration (Foster 1999). This rift is a tear in a social and natural metabo-
lism in which systems were aligned, which Marx observed as a basic instabil-
ity of capitalist production. This notion of the rift between farmers and their
crops, between urbanites and rural people, figures in research on human-plant
politics. The rift may help describe some of the influences of capitalist modes
of food production, urban design, and the obscuring of the supply chains and
origins of food and other products. Would one buy a piece of furniture if photos of a clear cut were hanging on the showroom wall? And yet these images
are important in a paradigm of treating plants as subjects.

MAPPING VEGETAL INFLUENCES

Across the various chapters, vegetal influences (as seen in the diagram below)
are shaping aspects of animal coexistence, politics, cognition, orientation,
identity, belonging, and place. Plants and human-plant relations in a framework
of mutual subjectivity open up convivial relations across animal species, as is
evident in the first chapter's case study about human-elephant conflict. Each of
these spheres in the diagram "engenders" the others. Forests engender learning
and politics in some worldviews, and the complex, situated relations in place
maintain identities, belongings, and resilient systems for humans and elephants.
An eye to these relational movements between the vegetal, the human, the
political, the land, place, emotion, and learning opens up alternatives to what
Latour (2018) calls the modernist taming and rationalizing of the Earth. The
climate crisis reveals the folly of these approaches; sharing the living world
with others has certain logics that differ from modernity's logics (figure 0.1).

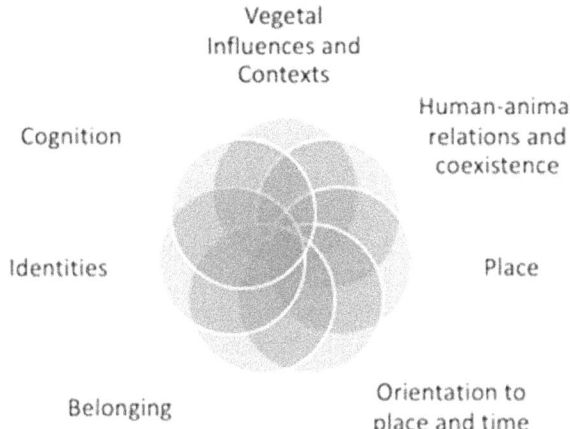

Figure 0.1   **Social and Subjective Domains Where Vegetal Influence Is Apparent in the Five Case Studies.** Elizabeth Oriel.

Across the case studies, the glue of phyto-human social worlds, the ties that bind can be found in intersubjectivity, affective ecologies, and poiesis. These are potent yet invisible forces of influence and connection. Intersubjectivity has been described as an "experiential sharing that occurs among subjects" (Reuther 2014:1001). Philosopher Edmund Husserl described this as shared meanings across subjects, while shared rhythms as a "manner of handling the world" is Maurice Merleau-Ponty's (2012:370) approach in a cross-fertilization of subjects. While shared meanings across humans and plants without a shared language may seem impossible, Merleau-Ponty's shared seasonal and daily rhythms are certainly at play in phyto-human relations such as in cultivation. Research among Indigenous communities speaks of intersubjectivities between humans and plants that are culturally significant (Dev 2018; Mentore 2012).

Affective ecologies, another aspect of this sociality, as developed by Neera Singh (2018), refers to the ways each being, body, and force in an ecosystem affects others. Drawing on Deleuze and Guattari, Spinoza, and Massumi, this approach to the living world de-centers humans and emphasizes how each is impacted by others, engaging with relationships, and becomings-with rather than distinct beings. Affect brings diverse agencies to light. In chapter 2, the jacaranda tree in Australia is presented as a place maker in terms of affective ecologies and diverse roles. Vegetal lives and their influences open up relations to context; in chapter 1, this is evident in how vegetal communities, practices in cultivation, ontologies, and identities in relation to the vegetal shape human-elephant coexistence. Each chapter takes a different approach to this social milieu.

Human-plant relations are explored in anthropology, cultural geography, and political ecology, as in how plants, when viewed and treated as inert resources, as things, contribute to empire building. A primary and popular text for this explication is Amitav Ghosh's (2021) *The Nutmeg's Curse*, which puts plants, colonialism, capitalism, politics, and earth-based logics into conversation, revealing how each societal crisis is interdependent with others, stemming from an erroneous perception of the earth as inanimate. Colonial rule, wealth, and empires were achieved through certain thing-like perceptions of and relations with plants, such as in sugarcane plantations and coffee and tea production in Sri Lanka and elsewhere in monocrops. The concept, Plantationocene, centers plantations with all attendant labor divisions, slavery, extraction, monocropping, and monologues as a primary organizing force of modernity and a source of current ecological breakdown by denying diverse species, authorities, knowledge, and relationships (Haraway 2015). Plants become a centrifugal force in this perspective, such that how plants are treated correlates with the treatment of others.

Vegetal life has co-created a habitable world for others, and yet plant monocultures are forming outside the plantation with so-called invasive species that are changing community structures worldwide. As Maan Barua (2023) suggests, vegetal agency can be pernicious, as with the *Mikania* vine in India, that was introduced from South America as a cover crop for tea plantations. The logic of plantations becomes an exponential spreader of extractive logics across landscapes, as plants move and "invade." In their power to both generate abundant life and to decimate biodiversity in monocultures, they hold great influence. The difference between biodiversity generation and erasure is the situated plant, in relations of co-evolution with its context. And such strong influence and sway could become central to land use decision-making. Permaculture, drawing on Indigenous knowledges, recognizes vegetal roles and influence and designs land with this in mind.

As described above, trees in numerous cultures are central to origin stories and thus orient humans to timelines and what came before. Trees have thus helped map and inform humans' subjective sense of self in relation to where they come from, and this is a cartography of vegetal subjectivity. The Norse Yggdrasil is a tree that holds the Nine Worlds, one of which is where humans reside. The tree forms a geography of existence and metaphysics (see more in chapter 5). An example of vegetal subjective geographies and Indigenous plant geography is *seya ania,* or Flower World, which for the Yaqui in the southwestern United States and Mexico, exists below the dawn, filled with brilliant colors where Little Brother Deer lives (Spicer 1980). *Seya ania* is one of at least nine worlds that overlap, such as *yo ania* or enchanted world, *tuka ania* or night world, *tenku ania* or dream world, *huya ania* or wilderness world. Flowers for the Yaqui are manifestations of souls, as well as metaphors for all that is good and beautiful. Humans can access *seya ania* (Flower World) through doing the deer dance. Flowers are also connected intricately with sound in song, in the same way that "chromatic" can mean sound and also color.

Humans have been influenced by and mapped the world according to plants and their presence and influences for millennia, in the trade of wapato roots in the Pacific Northwest (chapter 3), for spices like nutmeg and cinnamon, and in botanical explorations such as those of Alexander von Humboldt. Humans have been carrying their culturally significant plant species around the world forever. Greeks carried their vines, Romans their wheat, Arabs their cotton, and Toltecs their maize (Von Humboldt 2006). Humans cut forests and drained wetlands in Europe, which created wholly different plant communities, dominated less by primitive species that reproduce asexually with spores, which made way for more angiosperms. The way humans have carried and cultivated plants as they travel coincides with a sense of plants and their placements as human-driven, without vegetal agency. And yet, an alternative story is just as likely, that plants domesticated humans and not the other way around (Pollan 2001). Humans may be a vehicle for plant agencies in their quest to evolve, as in the advancement of wheat species through agricultural modifications (Head and Atchison 2009). Perhaps plants in some sense orchestrate these botanical movements.

And yet, outside of these trade routes, plants have their own geographies that have created habitability through relationships across the globe. Von Humboldt's volume, *Views of Nature*, is considered the first ecological book (Wulf 2015), expressing relationships as primary to nature, explaining how each being's lifeways support the whole system. Humboldt created a botanical map, *Ein Naturgemälde der Anden*, depicting plant locations and measurements such as altitude, atmospheric pressure, temperature, and humidity up the slopes of Chimborazo, a volcano in Ecuador. He did not make it to the summit but climbed 19,286 feet, which helped settle his explorations into a perspective. From this emerged his concept of nature as a global force with interrelations across topography, climate, volcanic activity, that helps explain plants' locations (Wulf 2015). Humboldt's map is an example of how plants map the land (figure 0.2).

This focus on relationships and land correlates with the concept of place as opposed to space or land. Von Humboldt viewed living beings as participants in human cultures and thus rejected Cartesian lines between nature and culture and nature as a machine. Machines are made up of parts that can be taken apart; thus, they are combinations of parts, while a living being is different in that the whole is greater than the sum of its parts and cannot be taken apart and put together again.

Maps used to be quite different. In medieval times, maps contained not only sites and place names but also stories, experiences, and associations.

Figure 0.2   Alexander von Humboldt's Tableau Physique, 1807.

Relationships to place have been hewn down to very narrow ideas of generic and objective truth about location and place. This obscures how place is co-created across diverse subjective beings with story and history as central features. Humans and other animals also possess internal mental maps of place, which orient them in movements and in meeting needs. Plants are central to both mental concepts of place, to experiences of place, and also to aspects of spatial organization and to stories-in-place. In rural west coastal Ireland, when I asked for directions, a local told me to look for a certain kind of tree to mark the right way. "When you get to the large pines, turn left." On my walks every day, I stay on track by noticing familiar trees as I go or groupings of trees. Plants are living landscape features that direct one on their way.

This association of trees, place, orientation, and mental maps calls to mind the mnemonic device called the memory palace. These are virtual spaces in one's imagination where knowledge and memory are stored and connected to certain locations. For example, sections of a speech are remembered by connecting each to various locations in a room. The effectiveness of this mnemonic exercise speaks to a spatial quality of memory. Memory palaces speak to the broader realm of cognition (as memory is central to thought), having place-based and spatial components. Trees and plants have structured the world, and many orient themselves to their route from familiar trees. Trees and plants are part of cognitive and memory markers, highlighting their influences on cognition and place-based knowledge. Research that reveals associations between trees and a human sense of belonging to place (Head and Muir 2005) may be accounted for in part from these lines of connection drawn across the spatial and the cognitive. This is taken up in chapter 2 on jacaranda trees and in the final synthesis (chapter 6).

Another example of how plants influence and map place is found in my friend Terra's garden in Portland, Oregon. She fostered an agroforest in her urban backyard, though in the front and parts of the backyard, she let plants (weeds) seed and root according to their own designs. According to Indigenous traditional knowledge as well as permaculture (which draws on Indigenous knowledge), plants that are called weeds actually seed and grow in places where their particular qualities benefit the soil, hydrology, mineral content, and generally improve relations-in-place. For example, in Terra's yard, dandelions with their deep tap roots break up degraded and compacted soils; comfrey brings minerals from deep down into surface soils, which benefits other plants; mullein brings moisture to overly dry soils. All of these species improve conditions and habitability, thus promoting life.

Plants' ecological roles also relate to their medicinal qualities. Dandelion breaks up toxicities in the liver, comfrey offers minerals and also clears toxins from the body, while mullein brings cooling and soothing to red and irritated respiratory tissues. In *The Ecology of Herbal Medicine*, Saville and Hardin (2021) write that plants' medicinal qualities produce similar effects on the land as they do in the human body. Volatile organic compounds in plants move energy in the body, and they have similar effects on land. Thus, plants are medicine to the land and soil. These benefits are highlighted in permaculture practices and yet have been known for millennia by Indigenous people and appear in plant stories. These are a few examples of plant influences and subjectivities, and each chapter takes a deeper dive into this terrain.

In chapter 1, plant influence mediates human relations with elephants in Sri Lanka. Human-plant relations and plants themselves are shown to be mediators of, and foundational to, human-elephant coexistence and conflict. A mental map of my fieldwork site in Sri Lanka from a colonial or developmental perspective views extractive potential from soils and places, from water bodies and centers of economic value. An alternate format is found in older subsistence farmers' mental maps across land and beings. These maps include histories of land use before the national park existed, interactions with elephants, poems recited to plants and animals, seasonal crops, Buddhist temples and their outreach, and small ancient human-built water bodies called tanks that irrigate the land in small areas and where elephants drink in dry seasons. For some of the farmers I met, their maps include social relations across beings and also express aspects of local autonomies, which allow local people to thrive without dependencies on distant economies. Relations to plants are primary in sharing landscapes.

In chapter 2, the influence of the non-native jacaranda trees on Australian culture and society is apparent through news articles across 123 years. Plant socialities and influences as place makers are so robust that colonial praxis involves re-making places in the image of both a familiar and idealized place by settlers who are displacing Indigenous people and also native plant species. A map of jacaranda influence in Australia through affect reveals emotional ties of belonging and identity to jacaranda blooming events, as well as a sense of time as dictated by the trees' blooming cycles.

In chapter 3, the plant wapato is central to cultural and placemaking for Indigenous people in the Pacific Northwest of North America. Working with biogeography and psychogeography, this chapter explores the impacts and influences of wapato over time in abundance of food, which shifted lifestyles toward being settled and sedentary. Wapato, a perennial aquatic plant with an edible tuber, shaped the place and terrain on what is now Sauvie Island before white settlers arrived. A shift away from wapato as primary in human diets altered the current floodplain. Wapato was central in trade and was an expression of the region's biogeography, with floodplains and volcanic nutrients making soils and plants especially fertile. This chapter is an example of how to conceive of place through vegetal actors, as they emerge and express forces such as climate and geology.

Willow is the plant actor in chapter 4, and I trace willow's influences on and with basket weavers in Denmark. Weavers grow their own willow, harvest, sort, dry, soak, and weave across different times of the year. This chapter foregrounds weavers' voices with long quotations in a polyphonic style that brings willow's own voice into the mix. Willow pulls each weaver into the vegetal world, attending to the plant's needs and interests, and disrupting lines of domestication. Plant influence can be mapped on the bodies of basket weavers in their gestures, in pulling and tugging on willow rods. Willow is the traditional plant for basket makers in northern Europe due to its strength, flexibility, and beauty, and these weavers carry on a millennia-long tradition. As one weaver says, "you have to be sensitive to the willow because willow has a will, and you have to feel it, and find a cooperation between you both." One could map willow cultivations and weaving groups as a form of bodily influence the plant exerts, as well as an embodied phyto-human sociality.

In chapter 5, ash tree influences and socialities are explored through extant literature and interviews as being both mythic and practical, making life easier in many ways while also being viewed as the World Tree that holds all together. Though ash tree geographies in the United States and Europe are shrinking from a feral insect and fungus that are fatal to trees, they have been central to relations to time as the origin of the first humans in some stories, and as the one to survive the end of time in Norse myths called Ragnarok. The World Tree carries a metaphysical geography that is present in many pre-industrial cultures. A dying World Tree is a potent mirror of the socio-ecological crisis for all beings with climate change and extinction. Vegetal lives, like the oaks on the Heath, are mirrors, guides, and offer an approach and orientation to learning from the world. Chapter 6 offers a synthesis and conclusion.

Each chapter presents a case study of vegetal influences that could stand alone, yet, when told all together, they reveal diverse local histories, stories, and ecological relations. The first several cases were developed independently, and at some point, bringing them together in a volume made sense. Plants come forward to upset monolithic approaches to experiencing place by offering layered relations with context, tracing how social and physical elements merge in emoting, thinking, orienting, and in aesthetics. The vegetal is a social, relational structure that influences animals, shapes how animals relate to one another, and shapes place. A central glue of human-plant relations is poiesis; bringing something new into the world, the aesthetics of which ground each to identities and belongings.

LEARNING WITH PLANTS IN RECURSION

Finally, vegetal influences have practical applications in a spatial map of learning and knowledge production through a reflexive process called recursion. Anthropologist Gregory Bateson wrote about recursion as central to learning, adaptation, self-organization, and evolution (Harries-Jones 1995). Partnering with plants that Myers invites with "planthropology" speaks to a hybrid form of learning with plants as participants, as shapers of the learning process. Attentiveness, sensitivity, and, as suggested earlier, alterations of humans' nervous systems may foster different perceptual and aesthetic modes. Plants actually instruct in how to learn about them through their structural forms and gestures. The spiral is a significant plant form, found in ancient fern species, sunflowers, cacti, pinecones, and more. This same spiral form occurs in mathematics in the Fibonacci sequence. Each new number that is added to the sequence is generated by adding the last two numbers together. Thus, the movement forward is instructed by looking back. This is a recursive process. "Recursion" comes from the Latin word *recursionem*, meaning "a running backward" or "return."

Recursive processes are self-referencing in the movement back and form spirals of movements into the world or system and back into the self or part of the system. Thus, they connect parts to the whole; they access patterns that make up the whole through feedback loops. Recursion helps to remove reifications in knowledge production; these are abstractions and concepts that do not reflect the living world and yet become more real to society. An example is the term "ecosystem," which has many definitions, and some argue that while being a primary word to describe nature, it is not a real or useful category as it does not reflect real-world designations. Reified words and concepts come to dictate relations and policies, and yet recursion that moves back in reflexive ways helps to prevent category errors and erroneous distinctions.

Recursion is also operative in aesthetics and in an aesthetic field that opens humans to their surroundings. Ecological aesthetics, Bateson notes, centers on habitats and perceiving changes in one's ecological surroundings (Harries-Jones 1995). In this way, recursion and aesthetics are approaches that open access to fuller perception and understanding of plant life. Plant life, for example, can be perceived as having similarities and differences with human life in a relational process of learning where one's own body, mind, and experiences are relevant to the recursive inquiry. Bateson takes his claim about aesthetics further by saying that the sacred is also invoked in recursion, as he defines the sacred as relations between parts and the whole, or between conscious and unconscious, or between poetry and prose.

In many of the chapters, I mention my own subjective experiences with the vegetal as a way to make my claims about vegetal influence with a circling back to self, as a reflexive lens that is in dialogue with one's understandings and arguments. Reflexive processes in education are finding wider appeal, and journaling and exercises that employ personal experience are widely used by teachers in various settings. Certainly, academic arguments are never free of subjective stances, and reflecting on their origin and nature in experience is important to strengthen transparency, while also highlighting the centrality of subjective experience to knowledge production.

NOTES

1. "Land base" is a term I adopt from California-based writer Derrick Jensen, referring to land along with human dependence on the land. It speaks to relational interdependence more than the term "environment."

2. See David Graeber and David Wengrow's *The Dawn of Everything: A New History of Humanity* (2021) for more on how Indigenous societies in North America inspired models of liberty and democracy for Enlightenment thinkers.

3. See Tim Ingold's (2017) work on meshworks and correspondence as relational becomings-with in the world across beings, place, and forces.

4. The geo-social is a term from Bruno Latour (Stein Pedersen, Latour, and Schultz 2019).

Chapter 1

How Vegetal Matters Shape Human-Elephant Relations and Coexistence

Plants are making their own policies.

Judy BlueHorse Skelton, 2020

Driving around southern Sri Lanka and inside Uda Walawe National Park, I see elephants' bones showing through their skin. Elephants are generally thin in the dry season, yet their condition suggests more than a seasonal issue. In the news are images of elephants standing on a landfill, eating trash (Mishra 2020), suggesting elephants have lost their terrain and are now dependent on human waste.

Elephants are exceeding the carrying capacity of protected areas (Fernando and Leimgruber 2011; Santiapillai et al. 2006). This means their land base cannot provide for them anymore. In many national parks, vegetation is changing, and what grows is not palatable to them. Meanwhile, their traditional movement routes have lost continuity as lands are developed and they lose access. Vegetative diversity in native plants has been decreasing, as the invasive species *Lantana camara,* or butterfly bush, grows in every direction, filling the air with a sickly-sweet perfume. Male elephants line the roadway at the southern edge of the park, standing adjacent to roadside stalls, where they beg for fruit from passersby.

In 2018 and 2019, my ethnographic fieldwork took me to numerous small-scale farms in southern Sri Lanka, near Uda Walawe National Park, to study human-elephant conflict (HEC). In the last decades, the conflict has been increasing, threatening elephant continuance and farmers' livelihoods. This interspecies territorial conflict occurs most often within cultivated fields where elephants enter to eat crops. I adopted a systems perspective[1], exploring how realities on the farm and village levels are shaped by and resist global forces in land designs and policies. On this island south of India, I wanted to understand how land use, which is often dictated by foreign entities in the last 70 years, has accelerated the conflict.

I explore here this interspecies conflict through the lens of plants' influences and how their relational and contextual geographies impact human-elephant relations. Early into my fieldwork, I gained a perspective that plants and human-plant relations were central to how humans and elephants related. A social world across plants, elephants, and farmers is evident in this terrain. Attending to the centrality of vegetal lives and relations, my research explored aspects of vegetal politics within forests and agricultural fields along with how practices mediate human-elephant relations, including conflict. Recognizing politics within relations to the vegetal is most common when considering human power dynamics, such as in colonial practices. A prime example is the British Empire gaining wealth and power through monoculture plantations of sugarcane in the Caribbean using slave labor.

Yet, the vegetal is more than a passive player and exerts a politics on the land. One example is how forests are spaces that foster diverse life forms and even autonomy for the many animals that rely on this vegetal community. Forests have long been viewed by colonial powers as spaces of hiding, as places of safety for rebels. In the Elizabethan time, the British cut down the forests in Ireland for timber to build ships; yet another motive was to prevent hiding places for those rebels who wanted to maintain their faith, language, and autonomy. Forests allow human and nonhuman animals to move undercover and to find food, along with all the ecological and cultural roles forests play. This autonomy from forest cover is one example of a subjective plant influence on animal minds and subjectivities that will be explored further in this chapter. The other chapters in this volume organize around one or two plants or one genus, while this chapter looks to myriad plant actors, including sugarcane, grasses such as paddy or rice, and butterfly bush among others.

Standing inside a sugarcane plantation, amidst tall stems that look anything but lush, I listen to a farmer describe his trials with the owning company and with the soil. Casting blame is common here. Though ironically, the Asian elephants who cause frequent crop damage are not blamed by farmers; numerous others are told off. The Forest Department, politicians, and Dole banana plantations are often blamed. A farmer shows me his banana trees torn apart from last night's elephant movements. Subsistence and small-scale farmers often lose one third of their crop to elephant damage, which equals their lifeline.

Elephants seek out crops amidst changes over the last 70 years in cultivation practices and hydrology, and amidst a dearth of their favored foraging plants. One could blame land development, overgrazing by water buffalo, introduced plant species that take over and are unpalatable to elephants, and their predilection for sweet fruits. Though killing elephants is both illegal and offensive to many farmers' tacit code of behavior toward other species, many elephants are dying. Explosives are hidden inside fruits called *hakka patas,* and electric fencing has voltage turned up to intentionally harm passing elephants.

Farming families in elephant-ranging areas have a different relationship with dusk now, as this is when elephants start moving and foraging, avoiding daytime when humans are active. Now, children fear walking outside their houses to use the outhouse, as bodily urges may lead to an elephant encounter in the dark. Schoolchildren in the early mornings fear walking to school and often come to school late. Elephants and farming families both need to be strategic and adaptive.

This blaming in the air is part of a fraught atmosphere in which farmers and elephants are tense, and little seems to be going right. The monsoons are changing their patterns, which has an immense effect on cultivation; overuse of pesticides has left many sick, in debt, and has left the soil unresponsive; families struggle with losing sleep to watch crops at night, and marriages start to unravel; many turn to alcohol to relieve their stress, yet this leads to a drunken boldness when encountering an elephant, which leads to more aggression from the pachyderms. The air is charged, as are the electric fences meant to keep elephants outside of crop areas.

In my field notes studying HEC in Sri Lanka, I have written the word "blame" over and over. When conflicts on the ground are so charged that bodies are restricted in movement, blaming becomes central as a release, as an attempt to understand, and as a communication style. Scapegoats, as Hannah Arendt argued, are imposed by many factors, one being ideologies. Here, the farmers I meet do not blame elephants or random scapegoats; they are quite grounded in their understanding of tracing the troubles and where they have emerged from. Many speak of the imposed mechanisms of change that have happened in the 400 years of colonial rule and postcolonial times with NGO development agendas and Western textbooks. And older farmers retain some knowledge of how water, topography, elephant thinking, and plants can mitigate problems. In other words, they listen to their land base.

Some farms I visited were two hectares, while plantations covered thousands of hectares. Small-scale independent farmers and those who work for others on plantations voiced the same concerns—*the soil was dead and elephant threats were rising.* The future seemed non-existent. Not surprisingly, in 2020, a year after my fieldwork ended, these same farmers rose up in large-scale protests and called for different land use plans that included elephants' lifeways as a priority. The lines of consequence, affect, and effect between farming styles and HEC were visible in these lands.

Plant geographies in this fraught terrain of escalating conflict are evident here and show their lines of influence. Over the last 50 years, decreasing forest cover that had extensive edge effects (forest edges are utilized by many species), successional changes in parks, increasing crop areas, and changing cultivation practices all have led to higher rates of HEC and reveal how access to terrain and thus vegetation (though water and minerals are also paramount) are central issues in human-elephant conflict. Thus, vegetal life is central to how humans and elephants coexist, or not. Accordingly, the interspecies conflict is mediated by human-plant and elephant-plant relationships. One could diagram this as a triad of humans-plants-elephants, with humans and elephants, both plant consumers, influenced by plants who are, in ecological terms, producers. The centrality of *socio-ecological relations with plants*, both wild and domesticated crops, offers a view of plants' enormous role in human-elephant relations and how relations with plants dictate animal relationships.

The Asian elephant in Sri Lanka is a forest species, yet thrives best in savannah/woodland areas and at forest edges where grass and secondary successional species grow (Sukumar 2011). Traditional slash-and-burn cultivation, or *chena* was a rainfed system, with growing seasons guided by monsoons. Chena also involved shifting crops, leaving fields fallow, which allowed for soil regeneration and allowed elephants to access harvest remnants. Yet, beyond human influences, elephants also play large roles, engineering the land through browsing, which kept areas in savannah-type plant communities, through seed dispersal, and through trail making through forests (Campos-Arceiz and Blake 2011). The forest openings maintained by elephants benefit humans with less closed canopy forests that weren't conducive to cultivation. In this way, relationalities with plants amidst both species benefited the other, with plants as a central ecological actor.

In the new scholarship on plant intelligence, communication, and agency (Gagliano 2015; Gagliano et al. 2014; Head et al. 2014; Trewavas 2016), plants are included in a widened circle of actors. If ecology is viewed as a political system across species, as Bruno Latour (2018) asserts, then all species are political beings. More-than-human political scholars have expanded the circle to elephants (de Silva and Srinivasan 2019; Evans and Adams 2018) and the field of vegetal politics expands politics to plants (Head et al. 2014; Marder 2013, 2012). Not only are other species in social relationship to humans (de Silva and Srinivasan 2019; Latour 2005), but agencies are multi-directional. As Barua (2016) writes, citing Michael Pollan's *Botany of Desire* (2001), plants domesticated humans as much as the other way around. Agencies and learnings reside in the spaces of relationalities (Barad 2007; Barua 2016).

Sri Lanka is a place to learn about sharing landscapes oriented around diverse logics, as ancient, traditional, and modern knowledge systems all coexist (though ancient ones are mostly remaining in the memories of the older generation). Young urban eco-friendly entrepreneurs I encountered in Colombo engage with all three knowledge systems. It was only in the 1960s and 1970s that cultivation and other practices scaled up with terraforming and technological mediation in agrochemicals and larger irrigated fields. Between the colonial and postcolonial dynamics, along with sophisticated ancient water systems that wet the dry zones during the dry seasons, and with modes of communities sharing land with large animals like elephants, this is a rich place to explore how imposed lifeways alter land-base relationships.

Plants in many traditional societies, in Sri Lanka and elsewhere, were recognized as knowledge holders, companions, and powers, not as things. Humans have communicated with them to ask for their help and assistance (Gagliano 2018; Van Daele 2008). As evolutionary biologist Monica Gagliano (2018) writes, colonial and neoliberal agricultural methods reproduce with plants the power dynamics within each system. In contrast to the "being-to-being" relations of traditional societies is the "being-to-thing" relation in modernist ways. In an expansive political landscape such as that held by traditional societies, ecological patterns of mutualisms, and reciprocal cycles that sustain systems are forms of *political action*. Simard (2021) writes about plant communities as models of republican, equitable resource sharing.

In this chapter, I explore plant-human relations and plant influences and geographies that mediate human-elephant relations regarding coexistence and conflict through (1) vegetative contexts and habitats (physicalities), 2) practices in cultivation, (3) rituals and sacred connections (identities), and (4) ontologies. The spectrum of relations within these categories forms interspecies social spheres. Yet, socio-ecological complexities make categorization difficult such that overlap exists across the sections. I draw on field research conducted during 2018–2019 in both the southeastern Uda Walawe region and the central Matale region near Knuckles Conservation Forest. Fifty interviews with diverse stakeholders across the two regions and in Colombo were conducted as part of more-than-human ethnography (Barua 2023) to access diverse human agencies and perspectives and those beyond the human.

PHYSICALITIES IN VEGETATIVE CONTEXTS

A natural place to explore human-plant relations and plant influences begins with the physical realities of plant life on the ground. The types of vegetative communities in a region form the matrix, the architecture of life, and set the tone of relationality across humans and elephants. Plants and trees provide sustenance for most life forms and circulate water through the earth and sky.

In this section, I explore how plant communities foster or minimize conviviality across species, shaping social relations.

Most interviewees blame the worsening conflict on elephants' lack of appropriate and sufficient vegetation to eat. One farmer on the edge of Uda Walawe National Park (UWNP) said that recently, elephants come and eat his chili peppers; elephants dislike chilis, which indicates to him that they don't have enough to eat in the national park. Elephants can be highly adaptive when it comes to their relationships with plants. However, the extent of forest cover is a huge feature for elephant lifeways.

The extent and continuity of forest cover are significant mediating factors in interspecies conflict, as Chartier et al. (2011) found in Assam, India. Their research using remote sensing data and social surveys suggests that with forest cover under 30–40%, HEC increases. In their study area in Assam, total forest cover, which refers to non-degraded forest types, decreased from 43% in 1973 to 37% in 1987, and many report that after 1982, HEC increased; deforestation was the largest land use change (Chartier et al. 2011:530). In Sri Lanka, forest cover was 28.7% in 2010, and from 2010–2019, the dry zone (primary elephant habitat) lost 8% of forest cover (Ranagalage et al. 2020:2). The extent of forest cover percentages varies based on the types of forest community counted, such as whether tree plantations are included; thus, these percentages are rooted in discourses about what constitutes forests. Vegetative categories play an enormous role in conflict resolution, as tree plantations are much less hospitable sites for elephants than naturally occurring woodlands (Kumar et al. 2010).

The findings that forest cover reduces conflict suggest a *connection between forest communities and elephants' abilities to meet their own needs outside of crops, and thus maintain some autonomy*, reducing dependencies on humans. Another study across 15,000 km^2 in northwestern Zimbabwe that is relevant to the Sri Lankan context finds that a threshold occurs when agriculturally transformed spaces dominate the landscape, increasing conflict (Hoare and Du Toit 1999). *Thus, the type and extent of plant communities across landscapes dictate and mediate other relationships.*

Asian elephants in Sri Lanka thrive in savannah/woodland habitats, which support 3–5 elephants per square kilometer, while only 0.2 elephants per square kilometer are supported in rainforests (Sukumar 2006). They graze on grasses and herbs and browse on taller vegetation in seasonally dry forests, which are *one of the most threatened forest ecosystems globally* (Fernando and Leimgruber 2011; Leimgruber et al. 2003). Being a forest edge species, elephants had plenty of habitat before colonial landscape changes due to shifting cultivation[2] practices that caused forest disturbances and maintained secondary successional growth (Fernando 2000). The large dams, reservoirs, plantations, and protected areas erected in the last 70 years have reduced landscape heterogeneity and forest edges, thus reducing suitable elephant habitat.

The land matrices in both of my fieldwork regions have altered in the last 70 years, with many more plantations and much reduced native plants and forest cover. In both areas, the presence of protected areas has mixed results for biocultural and multispecies relations. The fortress conservation model of separate spaces for wildlife is troubled by many (de Silva and Srinivasan 2019; Kamau and Sluyter 2018), and my own research bears this out. Protected areas generally ban any cultivation, and the lack of cultivation creates a lack of early successional plant communities due to a lack of ecological disturbance. This, combined with illegal grazing in protected areas by domestic/feral livestock, leads to decreased grass and herb biomass that elephants thrive on inside protected areas. While space for wildlife and naturally occurring plant communities is viewed by most local farmer interviewees as important, illegal encroachments into a protected area with poaching, grazing, and cannabis cultivation increase human-elephant conflicts.

Also, without *chena* cultivation and with illegal grazing, invasive plant species such as *Lantana camara* and *Phyllanthus polyphyllus* have multiplied. Anecdotally, in a jeep drive through UWNP, the sickly-sweet scents of *Lantana* overtake all others. *Lantana* reduces native browse species that elephants rely on, and is also unpalatable to them (though one interviewee suggested that elephants may eventually adapt to eat them.)

Plants, when removed from their ecological relations, can reproduce colonial and neoliberal approaches by taking over, crowding out native species that animals and insects rely on, and are often unpalatable to many species. *Lantana camara,* or butterfly bush, is native to Central and South America and was introduced to a botanic garden in Calcutta by the British in 1809 (Thakur et al. 1992). In South Asia, Lantana went feral, spreading especially in national parks where pressures from illegal grazing had decimated native vegetation. Lantana has allelopathic properties, which means it produces and then introduces biochemicals into the local soil that inhibit the growth and reproduction of other species (Priyanka and Joshi 2013). With strong root systems, they grow successfully in dense thickets in varied environments and thus alter local ecologies significantly. Successional processes are altered, reducing native species richness, increasing the size of wildfires, while also harboring malarial mosquitoes that are harmful or lethal to animals that consume its leaves and flowers.

This invasive species reproduces colonial effects and practices, reducing biodiversity and carrying a universalizing or centralizing effect on ecosystems. They disrupt the thriving of diverse agencies and transform naturally occurring ecosystems into something akin to a sugar plantation, where one plant species is in charge and dominant. Landscapes become less food-secure, necessitating the need to move farther to find vegetation. The rhythms of diverse plant species flowering and dying back offer pollinators and decomposers food in staggered succession, which is lost with the invasive species' monopolies, posing starvation threats for many. This invasive plant mirrors some of the human-imposed frameworks of centrism, a kind of universalism, in which all have to fit into one imposed framework. In this way, human-plant relations set the tone for how other relationships take place on the land.

National Park plant communities are degraded. At one time in the 1990s, 20,000 buffalo were grazing in the national park, owned by local ranchers. They needed more land to expand buffalo ranching beyond their own land's limits. This overgrazing caused the grassland communities to collapse in the park and increased conflict as grass is elephants' primary and preferred vegetation, one park warden told me. With no community involvement in protecting the park, which in other countries includes locals gaining revenue from the park,[3] there is little monitoring of these encroachments that destroy natural communities.

And yet, the cultivation of grasslands in the park is not recommended by many conservation officials, as wildlife needs human-free spaces to inhabit. These spaces, where tourists ride in fast-moving jeeps, are degraded spaces to some extent, with collapsed grasslands, invasive species, and illegal cultivation. The top-down management of protected areas, which extends a top-down approach to human-plant relations, where local farmers obtain none of the benefits from protected areas and have no power to monitor illegal use of the park, leads to greater divisions between farmers and wealthy elites who visit the park, and greater encroachment.

In the Knuckles area in central Sri Lanka, two recent land use changes are the 2010 UNESCO designation for the Knuckles forest, which led to a ban on *chena* and thus more closed canopy forests and no post-harvest browsing for elephants. The other change is the large Moragahakanda dam and reservoir, which provides irrigation for 82,000 hectares (Senanayake 2018) and coincides temporally with reported increasing rates of HEC. Farmers just south of Knuckles, near Udadumbara, are receiving UN support to grow cash crops and create bigger home gardens to offset these challenges.

Historically, this region is the northern edge of a grand swath of jungle, cut for tea and coffee plantations during the British colonial period beginning in 1867 until 1948, expelling elephants from the lush hills to the dry zone (Fernando 2000); the highlands still support many tea plantations. The plantation style of the human-plant relationship will be discussed in the next section, yet the physicality of monocultures, which excludes all other beings, is a monologue that is unlike how plant communities function in the pre-existing jungles, where resource boundaries are permeable and shared across diverse beings and roles.

PRACTICES, PLANTATIONS, AND POLITICS

In these fraught contexts, I came to see evidence that agricultural practices form the matrix of extensive landscapes and land uses. I also could draw lines of influence as this matrix of praxis sets the stage and the tenor of not only human-plant relations, but other relations, such as human-elephant. Settled, irrigated, and continuous cultivation with no shifting of fields and no fallow periods provides a challenging landscape context for human-elephant coexistence (Fernando et al. 2005; Thakur et al. 2016). In contrast, traditional shifting practices fostered coexistence in Sri Lanka and elsewhere (Fernando et al. 2005; Henle et al. 2008). The island's three traditional agricultural models—shifting, paddy cultivation, agroforestry are now, to varying degrees, infused with modern methods such as monocropping and chemical fertilizers, though most home gardens remain chemical free.

Each system fosters or limits plant interaction, communication, agency, and community structure. When farmers plant diverse crops amidst trees in traditional shifting cultivation called *chena* and in home gardens, plants have various and multiple ecological roles in preventing insects, providing nitrogen or minerals to the soil, providing food for birds, and communicating through fungal networks. Within industrial monocultures, plants are sprayed with chemical pesticides and are interacting less as mycorrhizal networks[4] are diminished; their complex ecologies are reduced and technologized to feed foreign markets. In traditional practices, soil health and the semi-natural setting allow plants to interact using underground fungal pathways (Gianinazzi et al. 2010) that act as communication networks (Gorzelak et al. 2015; Simard, 2018), while industrial methods limit these interconnections (Matson et al. 1997). Two factors—the *location and size* of agricultural fields and the *methods* employed—are both equally important. Industrial irrigated monocrops adjacent to protected areas increase conflict (Thakur et al. 2016), as crops are grown year-round, with no fallow period, and allow no access for elephants to browse when they leave the reserve. In this way, the practices that shape an interspecies sociality among humans and plants dictate animal relationships.

Agricultural production and irrigation systems have had a profound influence on elephant lives in Sri Lanka. As human civilization grew in Sri Lanka, elephants have moved into unsettled areas to avoid humans and gain access to food and water resources. The Sri Lankan landscape has dry regions, wet regions, and montane ecosystem types. Around 2500 years ago, ancient civilizations irrigated and planted in the dry zones with sophisticated systems of water tanks with engineered channels to move water. Aspects of these permanent agricultural systems excluded elephants, though in other areas, shifting agriculture (or slash-and-burn) combined with freshwater reservoirs allowed elephants to prosper (Fernando 2000). As mentioned previously, colonial rulers (1505–1948) and especially the British (1815–1948), developed landscapes into large-scale commercial production of coffee, tea, coconuts, and rubber. Starting before independence in 1948, and since then, dry zone irrigated agriculture has expanded, in large part through the Accelerated Mahaweli Development Project, which transformed 40% of the island's landscape with eight dams irrigating thousands of acres. These more newly established agricultural sites are where most HEC exists today (Prakash et al. 2020), though the conflict exists throughout elephant range areas.

Human-human politics and human-plant politics are mirrored systems. Human-plant relations involve power dynamics within the human realm, such as cultivator to land owner, corporate seed and chemical producers to cultivator; plant-plant community dynamics are also key players in agriculture, such as nitrogen-fixing and fungal communication amidst diverse plantings or monocrops with soil microorganisms and fungal pathways reduced. Aspects of plant propagation, such as seed sourcing, grafting, and cloning, are all studied as elements of human-plant relations with a prevailing politics (Fleming 2017).

Plant communities have an organizing logic, and this is both influenced by and influences the cultivator. In vegetal geographies and politics, the more plants exert their ecological roles, the more they engage politically as actors and less as instruments. Commodification, and yet also diversity, decentralization, and democratization are operative within human-human and human-plant politics. These logics, self-organizing principles, or politics extend to other species, such as elephants and other wildlife who enter cropped areas. Traditional shifting practices, which have been made obsolete by Western methods, fostered coexistence in Sri Lanka and elsewhere (Fernando et al. 2005; Henle et al. 2008).

Currently, farmers in dry zones contend with numerous challenges: wild animal attacks, drought, lack of stable incomes, rising production costs, competition with imported products, inconsistencies in land rights legislation, legal suits by the Forest Conservation Department, and improperly managed water systems. In these scenarios, farmers and elephants are both marginalized—farmers have less land to access, and elephants in small, protected areas lack adequate vegetative browse and water. In Yala National Park, over 50% of elephant calves have been starving to death due to a lack of adequate vegetation (Daily FT 2018). Farmers are drawn to industrial methods to earn more from cash crops; this shift increases dependence on corporate entities such as agrochemical companies, reduces soil and ecosystem resilience and health, and changes the human-plant relationship. This corporate hold on the process also exerts influence on plant agency and community structure. A central logic of industrial agriculture is control over natural processes for profit (Shiva 2016); a central logic of natural systems is to sustain the system through reciprocal relations, mutualisms, competition, and other dynamics (Berkes 2017).

Farmers' relations to plants within the ancient systems of *chena* and agroforestry were communicative, reciprocal ones in which farmers would honor local deities, ask for bountiful crops, and forgiveness for harming animals and plants with their methods (Gunasena and Pushpakumara 2015). Buddhist and Hindu farmers would practice the same rituals. Plants were spoken to, sang to (Ratnapala 1980), and were carefully protected for their ecological roles, enabling systems to thrive. *Chena* cultivation was a rain-fed, mixed-cropped, slash-and-burn system that was part of a larger village-level system that harnessed the logics of terrain, climate, soil health, and built-in practices to avoid elephant incursions. In dry uplands were fields of vegetables and some grains, while in the lowlands, paddy production took place. Forests were not burned indiscriminately; rather, farmers followed a strict ecologically informed system that prescribed which size trees to burn, protecting forests and allowing trees and vegetation to regenerate (Bandara 2007:3). Vegetation was also cleared with prescribed practices of cutting, pruning, and leaving shrubs and trees, which fostered forest regeneration.

Cultivation spaces worked with the logics of topography, sited at meetings of uplands and valleys, and spaces were planned based on function, with upland forests for water filtration, and plants and animals participating with various roles (Handawela 2016). Vegetation around the edges of paddy fields supported insects, was tended to and not cut, to support insect life and prevent crops from infestations (Bandara 2007:iv). Vegetation would be cut back in a field, though trees would remain, and after burning the fields, crops were planted—the dominant ones being maize, finger millet, sesame, green gram, black gram, chilies, and cowpeas (Gunasena and Pushpakumara 2015). Harvests would take place at different times and after two to three years, farmers would shift to a new field and repeat the process. *After crops were harvested at the end of rainy seasons, elephants and other species browse on crop remnants.* Syncopated relational rhythms across time and space, with elephants eating crop remnants amidst shifting fields, allowed for a choreography of coexistence. Human-plant relations were guided by seasons, monsoons, and strict regulations of use and practice.

In the colonial period, plants became commodities, instruments of wealth accumulation with the advent of plantations growing cash crops such as coffee, tea, rubber, and more. Changes in agricultural practices over the last 40 years in Sri Lanka can be characterized by technological intermediaries between humans and plants. Currently, *chena* farmers are planting more monocrops for cash, relying on chemical fertilizers and pesticides and poorly maintained water management systems and no longer shifting to new areas to regenerate soils. The soil requires enormous chemical inputs, farmers told me, and yet elephants come and eat one-third of their crops. The poor soil quality ties farmers to the increasing precarity of rainfall amounts due to climate change, to debt cycles for agrichemical purchases, and to less nutritious diets. The diversity of seeds, crops, and community support is diminished.

Continuous monocropping increases conflict with elephants as the lack of a fallow period means that elephants have no access to crop remnants after harvest. Wildlife scientists have recommended in numerous papers that *chena* be practiced adjacent to protected areas, as this system historically fosters a sharing of landscapes with elephants (Benadusi 2015; Fernando et al. 2005; Fernando 2015). The modern methods of monoculture, chemical use, and continuous growing with no shifting are *instrumental relationships* in which plants are things, instruments of wealth and capital accumulation.

The impacts of agrochemical use extend widely in terms of health issues and cycles of debt. As one prominent NGO director remarked in an interview:

> a lot of government agencies have promoted this model, that the farmer should learn from the fertilizer dealer. You go to the fertilizer dealer and ask to diagnose, ask what to apply, what chemical cocktails. You're putting the farmer and the environment in low priority, and you're not only creating a health mess but you're creating an import bill. And then you get sick people and then you import drugs. You have community issues, but if you draw the circle bigger, you have something else.

This larger circle of impacts, of agrochemicals and outside authorities, such as the Western industrial model of agriculture, is evident in the toll national debt takes from foreign direct investment. An economist I interviewed described the stasis from national debt:

> Our government doesn't have capacity. Due to our debt, we can't bargain with anyone. We can't even maintain the life of our people. We become very dependent people. If we don't have debt, we can design big projects, we can bargain.

Thus far, this section speaks mostly of the centrality of human designs and patterns in relation to the vegetal, and less of plant agencies. Plants exert widespread influences in their practices of producing habitability, with chemical scents that animals respond to, with the proliferation of invasive species, with fungal networks that transmit communications. In Uda Walawe, a sugarcane plantation on the southern edge of the National Park stretches 4,215 hectares. Smells of sugarcane (*Saccharum officinarum*) travel up past the park boundary, luring elephants to break through the fences and eat the sweet fibrous stems. Commercial crops of maize (*Zea mays*), gingelly (*Sesamum indicum*), banana (*Musa paradisiacal*), coconut (*Cocos nucifera*), cassava (*Manihot esculenta*) are grown at other boundaries of the park (Isthikar 2015). These species lure elephants (Schmitt et al. 2018), and this is one avenue of plants' mediation of animal relations across landscapes.

Elephants' olfactory sense is one of the most highly developed in mammals and contributes to social cohesion, as scent tells locations of others in female herds. Having a broad olfactory range depends on the number of smell receptors, and elephants have 2,000 genes for smell, while scent-oriented dogs have 811 (Schmitt et al. 2018). Elephants use scent to detect food, water, and also detect urine, feces, saliva, and temporal gland secretions. As grasses in UWNP are less abundant due to overgrazing, successional changes, and invasive species, crop smells lure elephants to leave the park boundaries, which are dangerous spaces as a number are injured by farmer attacks outside the park.

Plants also exert agencies in their interactions with their environment. A few primary plant actors in HEC are guinea grass (*Urochlia maxima*), butterfly bush (*Lantana camara*), jackfruit trees (*Artocarpus heterophyllus*), teak trees (*Tectonia grandis*), *Gliricidia sepium*, and the rice plant (*Oryza sativa*). Guinea grass is a non-native perennial grass originating from Africa, introduced to Sri Lanka in 1801 as fodder for horses and cows, and is considered an invasive species (Wisumperuma 2007); park rangers say the grass is beneficial to mitigating HEC as elephants prefer this grass species. *Lantana* is an invasive species that outcompetes many native plants and grows everywhere in UWNP; this outcompeting species increases HEC as elephants find the shrub unpalatable, and thus, vegetative resources inside the park are reduced.

Some of elephants' preferred leaves are jack and teak bark and young leaves; teak draws minerals and water from the soil and provides nutrient-dense browsing. Teak has complex impacts, as the wood is valuable and is grown in plantations around Uda Walawe, yet these trees are thirsty, reducing the water table. Glyricidia is a nitrogen-fixing tree that is often planted among tea and black pepper crops to enhance soil nutrients; this species improves soils and thus reduces chemical use, as the plant works in partnership with human cultivation. These plants all act in differing ways through their ecological roles, through relations with humans, and have various influences in HEC.

Intensification of landscapes, with crops that require large quantities of water, drawing minerals from the soil, exerts influences on wildlife such as elephants. Sugarcane, teak, and kaya, which cover extensive areas around UWNP and further east, all decrease water tables with their high water needs. The governmental Forest Department chooses tree species for their plantations, while other corporations and government agencies introduce sugarcane and banana trees in the area. These monocrop decisions may exert more of an influence on HEC than is acknowledged in the literature. Reduced water tables and depleted soils open room for opportunistic species like *Lantana* to outcompete native species. Plant diversity allows elephants to choose favored plants and also to self-medicate through eating certain species (Dubost et al. 2019; Shurkin 2014); this diversity is highly diminished in the Uda Walawe region where invasive species are ever-present (figures 1.1 and 1.2).

In increasing areas in Sri Lanka, plants are asked to play roles as intermediaries, as buffers that block elephants from entering a farm. Farmers plant biofences of agave and other prickly plants that deter elephants, as well as lime and orange trees since elephants dislike citrus. Regarding these vegetal intermediaries, farmers report varied results. Plants' scents and textures work in some cases as ecological diplomats, asking elephants to avoid decimating

Figure 1.1   Asian Elephants in Uda Walawe National Park, Eating Young Teak. Photo by Elizabeth Oriel.

Figure 1.2 Banana Tree Damaged by Elephants. Photo by Elizabeth Oriel.

crops. Another diplomat is the bee; in regions of Africa, bee hives are placed on farm perimeters, deterring elephants, as they dislike bees (King et al. 2011, 2007); research into Asian bees in Sri Lanka is ongoing (King et al. 2018). Farmers also proposed in interviews that the Forest Department, which manages national park buffer zones, plant elephant-appealing tree varieties to give them enough to eat before they exit the park boundary.

These practices that rely on plants and their relational roles are rooted in ecological processes and human-plant partnerships and plant agencies. With the loss of traditional cultivation that honored plants and the associated deities as respected beings, awareness of plants' agencies diminishes in younger generations, as the dominant Western worldview that has global influence views plants as things (Tsing 2012) or as invisible (Balding and Williams 2016); though strong plant relationalities are evident to me even among city dwellers in Sri Lanka who know many plant names and medicinal qualities.

As agriculture expands and intensifies, cultivation approaches become a primary organizing force for many landscapes, dominating ecosystems and relationships on the land. And yet, with new breakthroughs in understanding plant communication, awareness, and decision-making, agriculture is no longer a matter of manipulating inert objects. Raising plants' status as political beings brings their experience and intelligence into the mix, and in a multi-directional approach, suggests that plant community structure may influence human community structure and human-elephant coexistence.

VEGETAL LIVES AND ECOCULTURAL IDENITITES IN SRI LANKA

Sri Lankans have historically perceived plants, herbs, and trees as having a soul, having sentience (Ratnapala 1980:209). The Bo or Bodhi tree, growing near many temples, is central to Sri Lankan culture, as Buddhists make up 70% (Hindus are 12.6% and Muslims 9.7%). The 2,250-year-old Bo (fig) tree at the temple in Anuradhapura, supposedly planted by King Devanampiya-tissa, is visited by many worshipers on poya (full moon) days and at other times. The tree is closely tied to the Buddha and his enlightenment and rituals show gratitude; the tree is sacred, as the holy basil plant is to many Hindu people (figure 1.3).

"We have a unique culture for plant protection. We have a poem culture. We have a long history of plant protection using singing and poems," one agrarian researcher told me. Though these interspecies dialogues have diminished with the use of technology (agrochemicals) as a replacement for plant protection, some farmers, especially in more remote regions, still profess to communicate across species. This perception of plants and other species as communicating beings is part of most animist cultures, and has receded in Sri Lanka among modernist ways, though only in the last 40–50 years, a UN agrarian official reports. Older farmers still remember traditional methods and interspecies communications.

The process of paddy cultivation involved a strong identification with plants. The paddy plant (or rice) in Sri Lanka traditionally goes through stages of growth over time that for Sinhalese farmers, corresponded to human reproductive cycles (Van Daele 2008). In fact, traditionally, farmers in Sri Lanka shared hybrid and intimate identities with other beings, and maintained communications across diverse actors on and with landscapes.

> The Western or naturalist form of identification . . . is fairly different from the Buddhist and Hinduist forms of identification in Sri Lanka in which interior properties can be exchanged between human and non-human beings, such as deities, animals and plants. (Van Daele 2008:295)

Van Daele finds a temporal and epistemological breach in the Green Revolution and agrochemical use that has altered identifications, cycles, and perceptions of who the rice plant is in relation to the human and the sacred realms. This technological fix/breach is also present in the field of

Figure 1.3   A Bodhi Tree at a Temple in Anuradhapura. Photo by Elizabeth Oriel.

human-elephant relations, as traditional *chena* and relations of identification in cultivation have been altered with modern technological agriculture.

TIME AND SPACE

In traditional human-plant relationships, time and space were both organized by rhythms that were dictated not by the human realm but by the biospheric realm. The rhythm of monsoons and moon phases dictated planting cycles. Coordinated restrictions on plant use and plant burning were all subject to rhythms of use and rest. Elephants had access to fields after harvest but not during. These codes around access and prohibition were interspecies codes of conduct, oriented around annual climatic rhythms. Lefebvre highlights the centrality of rhythm to social relations and temporalities in Rhythmanalysis (2004). Intersubjectivity, or the ability to share experiences, is also about sharing rhythms, as Merleau-Ponty argued, and this was evident in the ways climatic rhythms were attended to as organizing principles.

Rhythm is how beings engage in temporalities and repetitions and is how individuals relate to the whole. Increasing conflicts can be attributed to rhythms that are discordant, out of sync, and also rhythms that are ignored or denied, as capitalism draws all into one universal purpose and timeframe. Endemic to rhythm is movement, and movement is central to all living beings, including elephants (Oriel and Frohoff 2020) and, less obviously, plants (Mancuso 2018). Western culture exerts a static approach to the earth, to living and being, which is out of sync with the seasons and cyclical temporalities.

In contrast, Sri Lankan and Indian history reveal complex socio-ecological relations among humans, plants, and elephants, interwoven with cosmo-ecological forces and beings. Each being has a role, a time, and space of access and retreat. The Buddhist reverence for all life, with prohibitions on killing other beings, and the Hindu pantheon of gods like Ganesh, who is both human and other animal, provide fields of ontological relations beyond the human that support shared spaces of cohabitation. Poya days in Sri Lanka are celebrations of the Moon/Sun/Earth's cycles, tying inhabitants to cyclical temporalities more than to linear ones.

The current ontology operating in expanding cropped fields in Sri Lanka is a modernist approach to plants. This is characterized by efforts at *scalability,* which is defined by Tsing as the ability to enlarge without distorting the frame (Tsing 2012). Tsing argues that scalability in plantation agriculture, especially in sugarcane that can be cloned, led to enormous expansion worldwide, avoiding the precarity of sexual reproduction. Thus, the plant and its scalability were instruments of imperial power. Scaling up becomes a way to intensify and also simplify the landscape, obscuring the heterogeneity necessary for living systems to prosper. Tsing says, "the living world is not amenable to precision-nested scales" (Tsing 2012:505) and yet scalability, expansion, and growth are central logics of imperialist and capitalist production within natural systems. In agricultural fields in Sri Lanka that move from multiple species to monocrops, scalability is at play in human-plant politics and impacts how farmers perceive plants and the ability of plants to exert their own decision-making processes. As one elephant researcher and former World Bank official stated, "it is irrigation and agriculture that increases human-elephant conflict." Continuous cropping alters relations to previous cycles and excludes other beings, such as elephants, from access to land.

Vandana Shiva's *Monocultures of the Mind* (1993) connects the modes and methods of growing plants in monocultures and the industrialization of living systems with a globalized homogeneity that crowds out diversity of culture, thought, and relationships. Shiva's work reveals how what we produce generates from and is intimately linked with how we perceive ourselves and the world. And this extends to relations with plants. A field of monoculture is designed with certain human priorities—to grow and sell one plant-based product, to crowd out other life, including soil microorganisms that cannot flourish with mechanized methods. This exclusionary, profit-driven design impacts all aspects of human society and human-animal relations. Shiva's work on biodemocracy highlights monocultures of both mind and agriculture, driven by capital and reductionist science, while advocating for a system of production built around the logic of diversity instead of uniformity.

DISCUSSION ABOUT PLANTS' ROLES AS MEDIATORS

As mentioned at the outset, elephants eating from landfills speak to dependencies and to landscapes that cannot accommodate the needs of wild elephants. The lack of wilderness, the close proximity of elephants to humans, and the ever-present cropped areas that attract elephants all lead some elephant experts to say that elephants are not wild anymore but are, to some extent, domesticated. They inhabit a human landscape with human foods, and have lost access to their usual wild plants, usual lifeways, and autonomies. The presence of naturally occurring forest at 30–40%, as mentioned above, provides habitat and context for species of all kinds to live in a wild state, free from human control and from humans shaping the integrity of their lives. Landscapes with reduced forest cover below 30% can be seen as landscapes of dependencies. Thus, a primary plant influence is in how forest communities provide spaces for species to be themselves, to possess autonomy. Plants and plant communities in this sense are the intermediaries, providing the structure for others' autonomies. This is a primary aspect of plant geographies and their abilities to influence animal subjectivities, and speaks to the profound and powerful role plants play in the lives of all species.

Beyond autonomies and conflict reduction, plant communities and relations among humans and plants mirror other relationalities. As discussed in the Practices section, when humans partner with plants and each has roles to perform in service to the whole, this dialogic approach serves multiplicitous interests. The roles are situated in ecological contexts and in rapport with the whole, have built-in limits, and ethics of use and overuse. The poems and songs previously sung to plants to maintain relationality and honor their offerings of life-giving sustenance exemplified a poiesis, a communicative and more equalized approach, and a recognition of plants' sentience and deep identification with some plants, such as paddy in particular.

Ultimately, industrial agriculture as practiced today does not foster a shared space of sentience and shared identification. Land sharing or land sparing is a central debate concerning how agricultural production can feed the world and also allow for biodiversity (Kremen 2015); should land be set aside for nonhumans, or should farms conform to conservation standards? The answer that this chapter argues for is that both are needed. Certain landscape features can counteract the negative effects of cultivation on wildlife, such as hedgerows and setting aside areas of non-production or land sparing (Phalan et al. 2011). Kremen argues that this debate reduces the complexity of the realities on the ground and reduces the options and that protected areas need to exist amidst favorable surrounding matrices.

Providing elephants with vegetative resources is one strategy to reduce conflict. One elephant researcher recommends that the Forest Department grow crops for elephants in the southern region. In Thailand, agroforestry projects include plants on the periphery for elephants to eat (Commons 2018). One plantation owner reported that he leaves paddy on the ground to prevent damage to his coconut trees. Alternatively, some farmers plant citrus, agave, and other prickly plants on outer boundaries that elephants tend to avoid as protection from crop invasion.

Climate change is expected to stress agricultural systems in Sri Lanka, causing significant yield declines projected by 2050 in rice and maize (De Silva et al. 2007; Nelson et al. 2009). The *Maha* and *Yala* are two annual monsoon periods in Sri Lanka; 66% of agriculture on the island relies on rainfall, which will be changing in quantity and quality in the coming decades, such that adaptation and mitigation planning will be important for food security. Sri Lankans are concerned about the coming changes, and more research is needed to understand food security in a holistic way, addressing impacts on crops but also on distribution and accessibility (Esham et al. 2017).

One agricultural practice to offset the changes in rainfall and temperature is small-scale agroforestry production, or home gardens (Weerahewa et al. 2012), which serve both as adaptation and mitigation to climate change. Home gardening is one of the oldest land use systems, second only to shifting agriculture, and involves a mixture of trees, woody perennials, annuals, and livestock. These are small-scale, multi-tier patches of vegetables, fruits, spices, medicinal herbs, and livestock, working with plant-plant and human-plant partnerships as the central logic of the system. While they are not a substitute for large-scale productions, home gardens have numerous benefits to farmers and others. They are also known in some areas of Asia to enhance coexistence among humans and elephants (Nyhus and Tilson 2004). Already covering 14% of land area in Sri Lanka, these systems offer farmers multiple and diverse harvests per year, require less physical labor than agriculture with annual plants, produce income, and are resilient against socio-economic and bio-physical changes on the landscape (Weerahewa et al. 2012).

Human-plant relations and politics mediate ethical relations across wider ecologies. This occurs in the forest cover and vegetative context that, within a certain percentage range, supports other beings, preventing human-elephant conflicts. The power dynamics, universalism, and anthropocentrism of mono-crops in which plants are objects, instruments for human consumption, undo these older pluriversal relationalities. The complexities of how plants use scent and how elephants interact with landscapes based on scent are central to land use designs oriented around cohabitation. Efforts to reduce HEC would benefit from the lens of human-plant relations. Partnership dynamics among humans and plants provide favorable matrices for human-elephant coexistence.

NOTES

1. See Donella Meadows for more on systems theory. Meadows (2008) describes systems as having three components—elements, interconnections, and purpose or function. Land-based practices are a system that organize around the purpose. In traditional practices, this was system resilience. The modern technological logic is extraction and, thus, profit.

2. Shifting cultivation is an ancient land use system in the tropics, and *chena* cultivation refers to the form in Sri Lanka. It has been defined as unirrigated, rainfed upland farming based on slash-and-burn, preferably of forest but also of shrub and grassland (Gunasena and Pushpakumara, 2015, p. 199). Fields were left fallow for 15–20 years, which allowed soil to regenerate, though this has been abandoned, which greatly reduces soil fertility (Gunasena and Pushpakumara, 2015).

3. In Nepal, there are farming communities that receive funds from park revenue and can enhance community services. This brings more local buy-in and monitoring to prevent illegal encroachment.

4. A mycorrhiza is a symbiosis between fungus and vascular plants' roots, involving the exchange of nutrients and diverse mutualisms, allowing plants to communicate about threats and more. Mycorrhizae improve plant and fungal fitness, and these networks enable forests to adapt to changes and be resilient (Gorzelak et al. 2015).

Chapter 2

Jacaranda Trees'
Affective Placemaking

As I begin writing this chapter, I visit a botanical garden in Denmark where I am living as a postdoctoral researcher. I sit down next to a *Jacaranda mimosifolia* tree in a section of the greenhouse with plants from the Americas. So many tiny delicate pointed leaves join together, almost fern-like, into one larger whole with fractal patterns; the bark is smooth and gray; branches hang in awkward arm-like poses. The feathery leaves remind me of fans, feathers, arms, gusts of wind, and fragmented light. Experiencing these trees in person is a priority, to have bodily, in-person contact. And yet, the setting, a botanic garden, which is so well attended and loved by the visitors I see around me, is a space of globalized uprooted beings, collected, commodified, researched, gene-edited, and more. Hardly akin to the mountainous ecosystems of the Andes where this species of jacaranda originates. My sense is that this tree is fairly healthy but not thriving.

The jacaranda tree, native to South and Central America and the West Indies, yet planted ornamentally on all continents (except Antarctica), inspires colonial imaginaries and outpourings of poetic verse, exerting influence as a placemaker. One of the almost 50 species, *Jacaranda mimosifolia*, commonly called "blue jacaranda," is native to the Andes mountains of Bolivia and Argentina, though it was planted in Australia starting in 1865. With purple-ish mauve trumpet-shaped blossoms that can last weeks to two months in springtime, they enact forms of vegetal (tree) influence on humans while also being objectified in colonial efforts to beautify and civilize; these complex relations exist in fields of placemaking and unmaking processes.

This chapter does not emerge from ethnographic research but takes on a different methodology in textual analysis, working with journalistic outputs as a way to view plant or vegetal influences on place and human relations. Here, I track the discourses related to jacaranda tree blooming cyclical events in Australian newspapers across 123 years (1900–2023), exploring complex multi-directional relationships that build place across vegetal affective fields and remake place in settler colonial processes. Contributing to Environmental Humanities' discussions of place, power, affect, and vegetal influence in Critical Plant Studies, this chapter uncovers how placemaking is a multispecies and affective process, and how the vegetal is a powerful force that is also objectified in settler discourses and processes of unmaking. Journalism has prominent placemaking roles as well, transforming spaces discursively into place[1] of meaning with social and cultural constructions (Gutsche 2014; Gutsche and Hess 2020); placemaking occurs both in human-plant relations and through the journalistic medium.

This analysis engages with a broadened concept of discourse in recognizing and speaking about plants and place. Discourse in the social sciences refers to meanings enacted through language and symbolic reference; yet with the relational turn, this definition has been enlarged to include meanings and communications from and with the living world (Abram 2017; Abram et al. 2020) such that trees can be said to engage with discourse within human-tree relational spaces. Discourse emerges from humans and other species, reflecting the nested arrangements in socio-ecological systems; to deny discourse to the living world is to erase this existential reality (Milstein and Castro-Sotomayor 2020), silencing other species' voices and responses to what industrial worlds are imposing on them. In this ecological crisis, how we speak about the living world is central to morality (Kohák 1984) and central to behaviors of recognition or denial.

Being in a relationship with plants in one's own surroundings is a form of placemaking and spatial orientation. "The backgrounding of plants is dangerous because it severs opportunities for dialogical interaction between humans and the environments in which they live" (Hall 2011:14). "We not only lose the ability to empathize and to see the non-human sphere in ethical terms, but also . . . get a false sense of our own character and location that includes an illusory sense of autonomy" (Plumwood 2002:9). A complexity of communicative modes surrounds the jacaranda in newspapers, with themes of urban and town beautifying projects, terra nullius, the effect of blue and purple flowers on humans, the globalization of plants, how certain grammatical constructions support more-than-human personhood, and more.

Jacaranda trees pose a formidable ecocultural presence in settler and other nations. With their springtime blooming, the trees become a phenomenon, inspiring festivals, tourism, poetry, and ceremonies. The cultural ties to jacaranda tree blooming echoes *hanami (*watching blossoms) in Japan, referring to cherry tree blooms, or *sakura*. For over a thousand years, the Japanese have engaged in *hanami in* daytime and *yozakura* in the nighttime; both involve picnics and family parties to enjoy the flowers. In Mexico, *Dicen las*

Jacaranda by Alberto Ruy-Sánchez (2019) is a poetry collection inspired by the experience of jacarandas, speaking of a collective utopia that he finds in trees' whispers. *Les Enfants de Jacaranda* by Sahara Delijani (2014) summons the jacaranda in another collective sense of belonging to a more just world, as a symbol and uniting force of those torn apart by political oppression in Iran. The Russian writer Vladimir Nabokov is known to have said he could live in Los Angeles for the jacarandas alone. While jacarandas have been planted in Australia, Mexico, the United States, Asia, and South Africa with similar dynamics, this chapter focuses on their presence in Australian newspapers and the human-vegetal relations on that continent.

Trees are themselves placemakers in their stability and structure, in how and what they offer to their surroundings. This is evident in Nabokov's statement about trees in Los Angeles. Trees tend to structure space for many; flowers are placemakers in their reproduction that involves insects and birds. Their scent and color in flowering that lures pollinators are affective qualities that generate the qualities and experiences of place. Angiosperms emerged 120 million years ago in the Cretaceous period, and the pollinator relationship is what allowed for bounteous diversity (Rose 2022).

> I find it mysteriously compelling that so much of what plants put forth to seduce nonhuman pollinators is seductive to humans as well: the scents of flowers, their colors and shapes, their timing. As is the case with many other manifestations of life, ancestral power is beautiful. (Rose 2022:140)

The shimmer that flowers give off is a lure and is transformational, carrying ancestral powers, Deborah Bird Rose writes, explaining Aboriginal perspectives and worldviews. Shimmer can be understood, perhaps, as a form of vegetal discourse, communicating to humans, insects, and others, and as a central aspect of phyto-human social worlds.

Trees not only nourish the soil and other species but also provide spatial orientation. Plants, in their lifeways, generate emplaced habitability for insects, animals, and fungi in symbiotic relationships such that the vegetal is primary to place and to diverse lifeways. These relational placemaking qualities lead to vegetal uses in colonial and settler processes of unmaking and remaking place. This chapter takes on these polarities of settler use of trees and trees own affective qualities in creating place, which intermingle and lose their distinctive boundaries in news articles, especially in the interwar decades.

Place is a category of thought and a constructed reality, anthropologist Arturo Escobar (2001) argues, and he writes that place, body, and environment integrate with each other. In a similar vein of dissolving divisions of body and mind, anthropologist Tim Ingold argues that embodiment is synonymous with enmindment. Place tells us "who and what we are by telling us where we are (and where we are not)" (Casey 1993:xv). Place has been described as space with cultural meaning (Cresswell 1996); it builds from deep experiences, feelings of familiarity, comfort, and connection (Massey 1991). Place is not static; it is vulnerable, pliable, involves action, and is where moral involvement matters (Forester 2021). "Many definitions of placemaking emphasize both the belonging aspects, such as sense of place, place-attachment, rootedness, etc., and the becoming aspects of collective reimagining/reinventing" (Barry and Agyeman 2020:24). One benefit of place as a conceptual frame is the focus on multi-directional affect between human and more-than-human lives and activities within a region, offering a systemic lens. Place can be an elusive concept in its relationality, yet highlights local autonomies and collectivities (Escobar 2020) and is essential to localized resilience and community cohesion (Hamdi 2010). Aboriginal people in Australia use the term "country" to refer to place and interdependent relations between beings and land. "Making peace with place" is what Deborah Bird Rose (2002) calls for among settler-descended people in Australia as part of an ethics of care.

And yet, "in these late modern times, the world has become increasingly placeless, a matter of mere sites instead of lived places, of sudden displacements instead of enduring implacements," (Casey 1993:xv). With place being so central to humans, the relations within place and placelessness, or *atopos*, are akin to relations between living and dying. Placelessness is a void, a desolation that begs one to dig in, and begin placemaking, and that emerges from colonial, settler, and modernist degradations of situated relationality. As Casey (1993) writes, philosophers across the Western spectrum describe the urge to fill up space with Being as a defense against *atopos*. Settler placemaking is first an unmaking, a blindness and undervaluing of emplaced relations, and attempts to eradicate human and more-than-human historical webs of relationships, an unraveling of country. For example, Grafton, Australia sits on the territory of the Bunjalung Nation, comprised of 15 tribal groups; these inhabitants, as well as the other species inhabiting place, are denied legitimacy in cultural, bodily, ecological, and biological ways.

SETTLER PROCESSES OF UN-MAKING AND RE-MAKING PLACE

Due to their centrality in human lives, lifeways, and to place, vegetal lives have been primary agents used to exert colonial and imperial projects. The entanglements of botany, plant collection, and empire have been well documented (Crosby and Worster 1988). Colonized land is perceived and conceived as empty space to be wielded for extraction; land is terra nullius (Moran and Berbary 2021), which allows for new plant assemblages like plantations, botanic garden collections, and non-native gardens to be placed in and fill the perceived emptiness.

One potent analytical lens engaging human-vegetal relations that deracinates place emerges in the Plantationocene. It refines the Anthropocene, highlighting how enclosed and extractive monocultures in plantations and attendant land use changes, along with enslaved labor, have re-ordered the world (Haraway 2015). Viewing biodiversity loss and the climate crisis as colonial legacies reveals that the novel ecosystems in which plants are moving, dying, and reorganizing due to rapid system change are also part of the threats of terra nullius. Land becomes generic sites and not places that orient us to the world, as in Casey's distinction. Many globalized plant species, sent around the world and grown in botanic gardens, colonize the land similarly to colonizers and settlers, becoming feral. This means they move outside the gardens and proliferate without the situated relations to place within ecological and evolutionary alliances and relationships that maintain proportionality as part of habitability. One example is the garden plant *Lantana camara* introduced in South Asia, whose proliferation is a cause of increasing human-elephant conflict (Oriel 2022; and see chapter 1 of this volume). Lupines in Iceland, brought in from the USA, are another example, with vast expanses of blue across the island in summer, crowding out native species.

Plants have been part of the calculus of nationalism in Australia and elsewhere, where they become signifiers of the nation-state (Ryan 2022) aiding unwittingly in forms of reification of nationhood. Nationalistic rhetoric uses analogies of plants and the land (Ryan 2022). Plants root into the soil, and this quality is drawn on to deliver nationalistic sentiment and a sense of belonging, and in Australia, it aided in justifying exclusions of Aboriginal inhabitants. The golden wattle tree became the national floral symbol in 1988 (Ryan 2022) solidifying settler identities with native plants and the Australian land base.

As a settler nation, Australian gardens and planted trees express colonial imperatives of domestication and Eurocentric ideals of beauty (Plumwood 2005). Settler gardens attempt to remake a place with an alien image and exert force on the land with deliberate feral species that need excessive chemicals to thrive, instead of collaborating with nature and place (Plumwood 2005). Planting jacarandas as street trees in Grafton, New South Wales, Australia has been an effort to civilize the city and is part of a larger movement to establish an Australian urban aesthetic (Frawley 2010). Settlers arrived in a land of extensive grasses that was shaped by Aboriginal fire practices; tree planting in this setting by tapping into the globalized network of plant bodies, plant knowledge that Sydney Botanic Garden maintained (Frawley

2010), was a form of settler unmaking and remaking of place through vegetal manipulations.

Grafton is the site of the Jacaranda Festival which began in 1935, and Grafton's newspaper figures prominently in the data for this chapter. Tree planting began in the 1870s with both native and non-native species, and in the 1900– to the 1920s, planting switched to only one species, *Jacaranda mimosifolia*. Street tree planting was in vogue across Europe and North America in the same decades (Frawley 2010). And yet, settler remaking processes intermingle with vegetal affects and influence, in twisting and overlapping branches of impact.

VEGETAL AFFECT AND AFFECTIVE FIELDS

As discussed above, place is space embedded with meanings, experiences, and connections. In fact, perceiving place or space as empty is a denial of others' lives and stories. Any space is actually someone's place. Place, Van Doren and Bird Rose argue, is relationally co-constituted across beings, place, and stories in entangled modes of intra-action (citing Barad 2007). Co-generating place can be understood through the lens of affect theory, such that place is a continual becoming through and with affective fields. While affect is defined differently across scholars, this chapter works with affect as how bodies affect each other, creating intensities (Stewart 2020), shaping mood, atmosphere, feeling, habits, identities, and lifeways. Bodies here include plant, human, water, discursive, political, among others. In terms of significance, the turn in the last decades to affectivity highlights how "living beings become who they are through reciprocity with that which affects and moves them" (Oele 2020). Vegetal affect is apparent across history and culture in plant stories, medicines, associations, and spiritual practices and realms. Plants are central to most cultures' origin stories, for example, which contain affective traces across time.

This chapter attends to how news articles express jacaranda trees' affective qualities, which have to do with atmosphere, color, cyclical repetitions, exotic imaginaries, dynamic influences, and shimmers that leave meaning in their wake. This chapter draws from Singh's affective ecologies (Singh 2018) as well as Stewart's and Oele's affect work. For Singh, affect is "a dynamic relationality between bodies of various kinds that enhances or diminishes the capacity of a body to affect or be affected" (2018:1), drawing on Deleuze. This affective relationality is not purely active or passive but moves in various ways, as each receives the world and responds.

Applying affective worlds to human-plant relations, Oele (2020) situates affect not with Deleuze and Spinoza's approach in embodiment but with

Aristotle, with an emphasis on experience and life-in-motion, among others. Oele situates affect within place, in a milieu, as this chapter does. For her, affect is a function of community co-emergence across beings and place, and ethics "should be responsive to the co-affectivity at the heart of community" (Oele 2020:7). Vegetal affectivity is described by Oele as a middle voice, which is a verbal form; this participatory voice does not separate the actor from the action (Ingold 2021) residing between the active and passive, and is present in Classical Greek and Sanskrit. The "agency of each being is actively present in their doing" (Ingold 2021:13). Oele emphasizes that the middle voice shifts "away from subjectivity and towards locality" (Oele 2020:22). This voice moves away from centralized selves as plants are not organized in centralized ways like animals are, toward states of acting and being acted upon. Plants contribute to structuring place and place structures animal lives, and in this sense, plants have a large role in shaping animal lives. One common thread in Critical Plant Studies is contending with how plants are conceived as inert, even though Darwin's *The Power of Movement in Plants* argues otherwise, and yet plant qualities have robust affective lines of movement that trace through place and imbue experiences with their affects; this is a real yet invisible form of movement.

To give some background on the tree in question, *Jacaranda mimosifolia*, is in the Bignoniaceae family and is native to and extant in the Piedmont forests in the Andes region of Bolivia and Argentina. They are declining in their native range, with an IUCN Vulnerable designation (Hills 2020), with agriculture, logging, and wood harvesting as primary threats. Piedmont forests are the most threatened forest type in Argentina (Brown 2006). Local uses of the wood include fires, timber for carpentry, and tool handles, and the bark is used as medicine for venereal diseases (Moostafa et al. 2014), while seeds and leaves treat liver and skin problems (Torrico et al. 1997). The tree is planted as a windbreak in agroforests, and cattle eat the leaves, leaf litter, and branches. This species grows to 25–50 feet, with bi-pinnately compound leaves and 2-inch long flowers on 12-inch panicles. Their bluish-purple color derives from anthocyanins, a pigment also found in sweet potatoes and black beans. Blossoms are trumpet-shaped and, when they fall, create a slippery goop on sidewalks from aphids feeding on flowers. The word "jacaranda" means fragrant in the Guarani language. Jacarandas are considered "invasive" in parts of South Africa and in Queensland, Australia.

Regarding methods, I worked with qualitative methods, synergizing discourse analysis with affective analysis. Australian newspaper articles were accessed through National Library of Australia archives online, while looking for articles on jacaranda trees, with special focus on the blooming period as this time period engages the most notice. Search terms included "jacaranda blooming," "flowering jacaranda," "blooming jacaranda," and "jacaranda tree." The number of articles using these search terms from decades starting in 1920 and ending in 1960 are in the thousands, while other decades have much less, and some decades have only 29 articles. Australia was chosen as the focal site because of the richness of news articles on jacarandas, and the settler relations to place. The sample size was 80 articles across the years 1900 to 2023, drawn from the online archive of newspapers held by the National Library of Australia. This time frame was chosen based on article accessibility in the archives, and because jacaranda trees were first planted in Australia in 1865. Purposive sampling allowed for article selection with subjective decisions on how relevant the articles were to the study. Articles were chosen for content about jacarandas that included cultural content related to the trees, while articles were disregarded if the tree was mentioned once, unrelated to the rest of the text. Purposive sampling was suitable as it did provide representative sampling and the goals of the study were not statistical or probability-focused, but were instead looking for trends in language and discourses.

Qualitative analysis involved coding the articles for themes, followed by interpretations rooted in discourse analysis and affective analysis (Wu 2018). Affective methods are challenging as affect is bodily, fleeting, and immaterial, necessitating inventiveness and experimentation (Knudsen and Stage 2015). This called for analyzing the texts for embodied qualities, felt experiences, moods, networks, and traces between them, and bodily rhythms (Knudsen and Stage 2015). These analytical methods involve uncovering meanings in complex relations that constitute place, including settler processes of both unmaking and remaking place. The analytic section that follows presents newspaper excerpts divided into settler colonial imaginaries, discourses, practices, and then vegetal affect and affective fields, and finally, climate change. Excerpts often contain examples of different discourses and affects, and not all are identified or discussed. One sentence may contain multi-directional influences from plants and humans.

SETTLER COLONIAL IMAGINARIES, DISCOURSES, AND PRACTICES

The analysis begins with newspaper excerpts that reveal discourses of settler imaginaries related to jacarandas as modes of relating to place in processes of unmaking and remaking.

> Even the small two trees in Jones Park are doing their bit, small as they are, to help bedeck our town . . . I would say that if you could see the town of Bourke bedecked as it is now by this glorious mauve blossom, your thoughts would at

once go back to those days of which I just spoke and you would try to picture Bourke of that time, as I often do, sunburnt and glaring, dusty and hot to walk in, ugly and bare, and you would again recall that except for a few old Coolabahs, which, God forgive us, we tried our hardest to eradicate, except for these and one or two fine old Morton Bay Figs and of course, the fine old Jacaranda at the courthouse, there are no trees to looks at. And, if you could just pop in now and see this beautiful scene as the Jacaranda takes over from the Western Australian Gums, you would hope as I do that this blessing that has been brought to this town on the Darling will never be jeopardised for the sake of a few pennies. (Nov. 8, 1968, *Western Herald*, "Jacaranda Time")

This excerpt describes Bourke, New South Wales, which is 800 km from Sydney, the capital, and is considered to be the gateway to the outback. Sitting on the Darling River and home to the Ngemba people, Bourke was surveyed in 1869 to establish a township. Indigenous inhabitants fought settler land theft and established a local reserve in 1946. The above excerpt typifies the settler discourse around terra nullius, or empty space. The sentiment is one of disgust for the way the land had been and pride in re-doing the place, eradicating the local and native species, which reverberates with Aboriginal removal from some local territories. From this local Bourke newspaper, the settler remaking discourse has human overtones, as the town is "bedecked" by two small jacaranda trees. This bedecking connotes discourses of pageantry and ornamentation as central to placemaking and is contrasted to the actual place as hot and dry, ugly and bare. The native species with long histories of co-evolution are *not* co-designers of place and thus need to go. Meanwhile, re-designing the land with designations of which vegetation can stay or go, and which exotics to replace natives with, is a process lacking attention to local conditions. The Morton Bay figs and the jacarandas are "fine," while Coolabahs and Western Australian gums are not. Coolabahs are a dry zone riparian species that germinate from flooding and provide significant riparian habitat with food, shelter, and shade for many species (Costelloe et al. 2016).

Bourke is a site of revision, as place is redefined and civilized (Frawley 2010) with jacarandas. This form of settler placemaking is "influenced by the twin forces of colonization and commodification, each of which selects in favor of exogenous ideals at odds with adaptation" (Plumwood 2005:1). Settlers set the aesthetic standards and, clearly as expressed here, devalue native species and aesthetics in favor of one that matches images from media or other colonized centers (Plumwood 2005). "The framing of particular plants as belonging or not in certain places is a culturally variable practice that pays only partial attention to the exuberance of planty life" (Head et al. 2014). This coincides with erasures of all local cultures, both human (Bacon 2019) and more-than-human, and denies trees and other beings as knowledge holders.

But the jacaranda tree I like best belongs to no home of ease and luxury. It grows in a sordid narrow street in Darlinghurst, in a backyard in the midst of a row of unpainted-depressed looking houses, all alike in dinginess and poverty. It is the only brave, beautiful note of color in the unlovely spot, and I think there are many who gain a message of hope and brightness as they look at the ethereal flowers which blossom so courageously in the ugly street. Many hearts will sorrow when the all-too-brief life is ended—till next November. (Dec. 4, 1922, *Evening News*, "Jacaranda")

From a Sydney-based news outlet, a jacaranda tree both generates and symbolizes upper-class luxuries and sensibilities, in contrast to poverty and squalor, asserting discourses of class difference that are legacies of colonial processes. Certain plants and animals in different cultures become symbols of wealth, luxury, and status, and thus participate in and even unwittingly assert class divisions. Beauty is a commodity in class structures, with upper-class urban spaces privileged with beautiful vegetation that lower-income groups lack access to experience. This privileging is echoed in Christian concepts of heaven as beautiful, clean, and pure. The jacaranda becomes such a symbol in the discourse of settler place remaking, and is appropriated as a commodity to serve this distinction. Related to the processes and discourses of civilized conqueror and uncivilized Indigenous inhabitants, this juxtaposition seems to interpret the tree's affect as privileged, as above the fray. The tree's beauty, one could argue, has the effect of making invisible the lack of equity or the underbelly of capitalist society. Somehow the tree is called "brave" and "courageous" to be in a "sordid" street, facing the realities of poverty. This correlation of jacarandas with wealth is echoed in other articles in the data set. The tree, as a settler, as a non-native, stands in contrast to poverty and to the local conditions, across both excerpts above.

The single jacaranda is a plant alone, without a community, which is a settler practice of de-centering community, removing, and devaluing contextual factors for all beings in settled spaces. Research that works with plant agencies takes seriously not the individual plant alone, but the plant communities that each plant resides in and evolves with (Elton 2021).

Harare's Jews, many of whom earn their living from manufacturing fabrics, clothing, and furniture, lead privileged lives. But there is a tenuous element to their lifestyles, since the political and economic future of the country is uncertain. Harare is a modern city, made beautiful by countless jacaranda trees covered with huge purple flowers that carpet its streets with purple. But outside the capital, people live in desperate conditions in mud huts, lacking proper housing, food, and medical care. (Nov 11, 1994, *Australian Jewish News,* "Zimbabwe Jews' concern for the future")

In this article from a Sydney-based Jewish newspaper, jacaranda trees in Zimbabwe are both beautifying, associated with privilege, and a discursive instrument for remaking place. This tree is cosmopolitan, circulating across continents in both material and discursive ways (Barua 2014) being a symbol and structure of empire in places. In this former British colony, most Jewish settlers were refugees from war and genocidal conditions, and yet here, jacarandas in the urban environment become associated with Jewish prosperity and privilege. The colonial and postcolonial demarcation line is clear: the beautiful purple-carpeted spaces are inside the line, and outside the line, people live desperate lives.

> The Queen Mother looked magnificent in a classic dress of jacaranda blue and a matching hat with osprey feathers. (Aug. 6, 1980, *The Australian Women's Weekly,* "Joyous 80th birthday thanksgiving for the beloved Queen Mother")

Here the bluish, purple color of jacaranda blossoms is appropriated by the Queen Mother, in the same way that the wattle tree became the Australian national symbol in 1988, becoming a discursive symbol of unity in nationhood within the Commonwealth, a body under foreign control. Clothing that mimics the local and beloved tree blossoms becomes a practice of empire and colonial control. While the settler discourses are blatant, the vegetal affective influences are also distinct, which is discussed next.

TREES AS AFFECTIVE BEINGS AND THE QUALITIES OF PLANT RECOGNITION

Affect theory, and affective ecologies in particular, position diverse humans, other species (including plants), and forces (like the wind and weather), all transmitting and contributing to a localized affective field. As such, plants are *subjects*, not objects, and this appears in linguistic forms, in grammar, and in content. Qualities are significant in this analysis, as they connect affect theory and conceptions of place. Casey (1993) describes place as having "character" and, as described above, affect involves qualities in mood, feeling, and atmosphere. Place, he writes, is also complementary with imagination and memory (Casey 1993:xvi). The affective qualities of trees in the news excerpts occur here in four modes: (1) in plants as subjects expressed in grammatical forms such as transitive clauses in which plants are the actors, being subjects of the sentence; (2) in tree qualities as perceived by humans; (3) in the powers or influences of trees and their qualities on human experience; and (4) finally, how the trees' qualities contribute to placemaking. The excerpts are listed chronologically to elucidate diachronic changes in discourses and trees' affective qualities.

A controversy has been carried on in Maryborough for some time as to whether the jacaranda tree is injurious to public health. It has been pointed out by some people that at the period of the year when the tree burst into bloom an epidemic occurs in the city, and this year there was a severe outbreak of influenza in Maryborough during the blossoming period of the trees. By a coincidence the jacaranda blooms during the dry period of the year, and the supporters of the jacaranda aver that the dry weather, dust, and bad drainage are the cause of the illness. (Dec. 7, 1918, *Kyogle Examiner*, "Jacaranda Bloom and Influenza")

Here in the Kyogle Examiner, based in Kyogle, New South Wales, the grammar and sentence structure give the trees no special subjectivity, but the jacaranda's influence is marked with the possibility of causing influenza. This article appears during the Great Influenza epidemic that claimed 40–50 million lives worldwide, though the article suggests this tree-illness correlation started before the epidemic. This excerpt reveals an affective power toward human bodies, related to the temporal co-emergence of blossoms and illness. The potency of the tree and blossoms is felt in this assertion, and many more articles cite this association from this time period.

A Jacaranda Tree

I sat within the house of prayer,
Untouched by ecstasy,
Though listing heaven's glories rare,
The good man's discourse—planned
with care!
No message brought to me!
he led me up no golden stair,
Beauty's High Priest to see!
When lo! Without, in summer glare,
A jacaranda tree,
A mystic thing—by angels kissed
to strange unearthly bloom,
Dreaming mid lovely lilac mist,
Softer than softest amethyst,
Frail blooms—too frail for earth I wist,
Dropped to their scented tomb!
I bowed in reverence—unaware,
And worshipped in the street,
Thus beauty built the golden stair
That led me to God's feet. By Emily Hemans Bulcock
(April 19, 1923, *Brisbane Courier*, "A Jacaranda Tree")

The jacaranda tree, here in the Brisbane Courier, exerts an effect of awe and mysticism, offering a genuine spiritual and mystical experience, contrasted with a lack thereof where it would normally be found, in church. The jacaranda is not an appropriated object but a divine subject, offering a physical experience within its purple amethyst bodies, both above in the branches and below on the sidewalks where blossoms collect. The jacaranda body is a place of experience, and in this instance, a mystical experience that is triggered from the tree's beauty, purple colors, along with something ineffable. Each of the four categories is present; although the tree is called a "thing," a sense of personhood emerges in the tree's powers and influence. As a placemaker, jacarandas here provide access to an imminent vegetal form of divinity of greater power than institutional religion offers. Interconnections of place and the divine appear in the word "Makon," which is a name for God in Hebrew, meaning place (Casey 1993:17).

The poem and affective power of jacaranda flowers is so potent and multidimensional that it correlates, as mentioned earlier, with the Aboriginal Yolngu term *bir'yun*, translated as shimmer, which anthropologist Bird Rose (2022) describes in her last book. *Bir'yun* or shimmer is a manifestation of ancestral power, which can be found emanating from flowers and is transformative. Bird Rose learned this from Aboriginal people in the Victoria River region in northern Australia and also draws on Morphy (1989), who describes shimmer for the Yolngu people of Australia as an aesthetic, an affective power, a sensory experience that can capture someone so that they can participate in ancestral power. Bird Rose (2022) speaks of shimmer in angiosperms, and specifically in flowers, as they entice and seduce nonhumans and humans alike. Shimmer arises in painting and ritual, dance, and song, and involves pulses of both dullness and brilliance.

As a painter builds up the surface with paint, which can be dull, this dullness is required for the next steps and for the potential transformative power of shimmer. All life carries this pulsing, taking place in seasons, in new growth. Shimmer speaks to some of the affective power of jacarandas in bloom, and how the trees bloom before their leaves come out, from a dull appearance to a brilliant one. Affect theory has also been called an "inventory of shimmers," a phrase from Roland Barthes, and carried forth by Seigworth and Gregg (2010). Shimmers in this context refer to the way affect in everyday encounters changes in intensities and, as Barthes says, can be noticeable in odors and luminosities.

Jacarandas! Jacarandas in every direction. They force themselves on your notice wherever you go; follow you into the office, chase through your mind all day, dispel your fatigue—if you will but dwell on their radiant glory—as they line your way home, and, at last, lull you to sleep by the memory of their swaying in the breeze, and pursue you in your dreams. (Oct. 31, 1926, *Sunday Mail*, "Jacaranda")

This passage from 1926 expresses all four of the above affective modes. Trees are subjects, actors who "force themselves on you . . . dispel your fatigue." In transitive clauses, the tree is the subject and the one who directs the action. The descriptive quality mentioned is their "radiant glory." Their influence comes as a power to "lull you to sleep . . . and pursue you in your dreams." These strong verbs and influencing movements speak to Oele and Ingold's work on the middle voice, in which the actor becomes their doing, or when being and doing synthesize. This article speaks of jacarandas as potent forces of character or qualities in place, as the trees' presence directs situated experience, actions, and feeling states (dispelling fatigue), as well as more ephemeral nighttime activities such as sleep and dreams.

> they are flaunting a blue-mauve glory against the sky. They lend a moment's colour to the drabness of the paling fence or the ugliness of the bottle-yard. From their position between the guarding figs and eucalyptus they wave one painted hand as if in invitation to rest beneath loveliness. They dare stand with their backs to the sea. Their colour puts the road signs to shame, and even makes the sky a little old and faded. (Nov. 22, 1930, *Sydney Morning Herald*, "Blue-Mauve")

Again, they are active and powerful, "flaunting," "waving one painted hand," and daring. Their influence here is in creating contrasts between the mundane, the drabness of the local place, "ugliness of the bottle-yard," with the beautiful and glorious. As a placemaker, jacarandas inspire utopian visions, chances to escape from the mundane qualities of native Australian landscapes and the drudgery of economic distress. This passage thus synergizes vegetal affect with settler remaking and denigration of local conditions that seem related to fears of and experiences of atopos.

> It is possible that in this year of 1942 with an enemy at the gates of Australia, we can appreciate beauty with an even keener tang, since a realisation of impermanency sharpens the edge of joy . . . Blue is the dominant note of Grafton now. It is adamant, boisterous in its demand for recognition. The trees are canopies of blue that in a very riot of prodigality shake a carrot of equal blue about their feet until the streets are inches deep in beauty. (Oct. 31, 1942, *Daily Examiner*, "Jacaranda Time")

In 1942, jacarandas are placemakers (in a Grafton newspaper) that antidote the misery of a world war, and the discourse relates to beauty signifying acceptance of impermanence. The trees gain recognition as subjects, not from grammatical construction, but the color blue is the active one here, being adamant and boisterous, suggesting strong affective qualities and character. Influence is in the beauty of the brief blooming period, and in the embodied physicality of the flowers that fall on the ground, so that human bodies and place become immersed in their affects. This excerpt, as do others, contrasts the vegetal world against the human industrial one, suggesting the vegetal is an access point to better, more beautiful, and more just possible worlds.

> First they speak in beauty. . . . They also speak of beauty. In an age dedicated to the ugly cult of atomic devastation, the message of beauty is of paramount importance. If we could transplant the beauty of the Jacarandas into every human heart the world would be nearer heaven in an instant, a poetic fancy. . . . Secondly, the jacarandas speak of power and purpose. They appear in their spring blossoms according to schedule. No meetings or reports or recommendations are necessary. They have a job to do and their power and glory appear each year on time. Just as well, or our Festival would never be held.
>
> From the jacarandas also comes a message of goodwill. The happiness of our famous festival is its chief appeal. For a few days the woes and worries of life are forgotten. (Nov. 5, 1953, *Daily Examiner,"* Jacaranda Days")

In this excerpt from the next decade (1953), jacarandas are subjects that have voice, communicating beauty, power, and purpose, which connotes personhood while asserting these qualities as necessary and needed by humans. Their affective influence appears to run parallel to the industrial world, as jacaranda worlds exert a discursive kind of social philosophy, an ontology rooted in beauty and goodwill. As placemakers, this influence creates spaces that counter war and nuclear weapons. They evoke a quality of essentialness, the primordial, or what Marder (2016) describes as the vegetal providing an essential ground of being.

> Each year in late spring, 3000 jacaranda trees in Grafton, NSW, burst into bloom and the whole city goes a little mad. For when the huge, glorious trees (allowed to grow taller than is usual in the capitals) spread their lavender-blue lace over Grafton and lay petalled carpet underfoot, the city celebrates the Jacaranda Festival. It erupts in jollity, as it has done annually for 41 years. This solid, dignified city on the Clarence River sees its steady citizens dress in shades of purple, place purple ribbons round the neck of dogs, while children ride on purple-painted bicycles which tone with the party hats, ice cream, streamers, and leis. (Dec. 3, 1975, *Australian Women's Weekly,* "Grafton's Glorious Jacaranda")

In the mid-1970s, the jacaranda trees continued to shape collective and cultural events with the springtime festival as shapers of place. They burst, spread, and lay—active verbal forms, connoting trees as subjects. The affective role of trees traces across to human experiences of jollity, to comportment of a gay kind, and to mimicking blossom color in clothing and all kinds of party accessories. The embodied experience of the trees as mystical, as communicating, is no longer highlighted in the 1970s, which coincides with the neoliberal era, globally deracinating cultural relations to place. The next excerpt is from five years ago and expresses a shift in an ethos toward vegetal lives.

> Jacarandas are an arboreal mirror that reflects the ugly state of our digital gratification-obsessed society. They're a short lived sugar hit of twigs and flowers that Instagram users inject directly into the social media main vein to live, laugh, love before the itch comes back and they move on to their next picture-perfect project.
>
> The trees themselves are all show and no go. The purple blooms that whip everyone into a frenzy last all of 30 seconds before the flowers fall off and sully the ground below with what can only be described as "moist spots." In a country built on hard work we openly celebrate a tree that spends most of the year doing nothing. (Oct. 3, 2018, *Sydney Morning Herald,* "The Jacaranda City")

Here in quite a strong contrast (2018), jacarandas are objects without their own subjectivity but are mirrors, reflecting society's and specifically social media's evils. Most jacaranda articles from 2020–2023 mention social media and the frenzy of taking selfies with jacaranda blossoms, reflecting corporate mediation of vegetal experiences. The trees' active qualities and influences are negative again, circling back to the 1918 article, though not with acute illness but with a chronic malaise of modern society and by sullying the ground with their blossoms. They are loafers here, contrasted with humans who built the nation "on hard work." The blossoming time is no longer an access point to other realities but is akin to a sugar rush, or a form of instant gratification that social media promulgates. The author portrays the jacaranda as an object that inspires frenzy, consistent with digital culture, in which seasons are commodified and objectified.

> As jacaranda trees begin to bloom, University of Queensland grounds manager Shane Biddell has warned students about a superstition on campus. "My wife, who was a UQ student, told me the myth she heard was that if you were hit by a falling jacaranda [blossom], you would fail your exam," he said. "I warn students now and tell them don't get hit, otherwise you'll fail your exam. But some tell me, no, it's the opposite—if you get hit, you'll ace it." (Sept. 27, 2020, *Brisbane Times,* "Blooming jacarandas set to cause havoc with UQ exam results")

Jacarandas' perceived affective roles are evident in how they influence those touched by the blossoms, leading to either exam failure or success. This superstition emerges from and with the trees' spatial and temporal positions, growing on university campuses and blooming during exams in springtime. This superstition travels with jacarandas, occurring in South Africa as well, and is somehow resonant with Amazonian lore where the tree is connected with wisdom and the moon.

A SYNTHESIS

These newspaper excerpts and analyses offer a sense of the complexity and intensity of human-plant social relations in the discursive domain of place, belonging, and worldmaking. For all the decades up until the 1970s, qualities of vegetal recognition and influence coexist with qualities of community and place and with settler remaking. Jacaranda influence appears in realms of bodily experience of the tree's shape, color, and beauty, which generate intimations of mysticism, creativity, beauty, pride of place, poiesis, and a vegetal temporal frame. These are phyto-situated discourses, with the human nested inside the vegetal structure of the world. Jacaranda trees offer affective and placemaking qualities in senses of home and belonging, in shared rituals of blossom festivals that echo through the history of cultural meaning in springtime flowers and rituals. This speaks to plants as ancestral, as living on Earth many eras before *Homo sapiens'* arrival, and recalls the Yolngu term *bir'yun*, or shimmer that runs through flowers as manifestations of ancestral powers (Rose 2022). The jacaranda tree in these articles is an affective agent that shapes the world, mirrors human conceptions and perceptions, and rattles at the divide of natureculture, especially before 1970. Yet processes of extraction, devaluing, and removal of localized ecocultures coexist with these vegetal influences, with efforts to remove local beings and to fill up and ornament the "empty," dusty landscape. One could argue that settler nations build their lifeworlds on top of the ground, often denying and attempting to discard embedded local histories and ecologies.

The nature/culture divide involves a depersonalization of nature and of humans from the living world that has two strands, according to Kohák (1984), one conceptual and one experiential. "In our time, however, the phenomenon has become global and the sense of depersonalisation of nature and of humans within it reaches far deeper" (1984:11).

> Since the seventeenth century, Western thought—and popular thought in its wake—gradually substituted a theoretical nature-construct for the nature of lived experience in the role of "reality" (1984:12) . . . .humans have to depersonalize their world in their imagination in order to be able to exploit it ruthlessly in their actions. (1984:11)

A complex personalization and also depersonalization occurs across the data set in accessing vegetal qualities and influence, with a loss of the living world as communicative. This echoes Raymond Williams (1976) on the word "nature" that he says holds the most complex meanings in the language, containing contradictions of what is most essential and also what is separate from humans, among others.

Journalism and the news media are significant sites for accessing news and information about the living world and other species, especially in this time of ecological crises. This chapter asserts how both the colonial and postcolonial discourses around place, as well as the vegetal affective and discursive qualities, can be accessed in journalism, a combination that is not well studied or documented. Journalism shapes public consciousness, and the potential for accessing vegetal subjectivities is substantial yet is likely unfavorable to corporate models of journalism.

A change is noticeable in the diachronic analysis around the 1960s and 1970s in language and conceptions of jacaranda trees. The transitive grammar that speaks of plants as persons is gone, as are the strong imaginaries into weaker ones. The land and vegetal inhabitants become depersonalized, as strangers in a strange land. This time period also sees the rise of neoliberalism, with Thatcher's haunting words, "there is no such thing as society," with deregulating public commons and infrastructure, leaving individuals to fend for themselves. In my work in Sri Lanka, a change from cultural landscapes to extractive landscapes, in which nature and culture are treated as separate spheres, took place in this same time frame (Oriel 2022). This new political economy extends to the vegetal, which becomes increasingly commodified, depersonalized, and perceived as mute objects.

Jacarandas and their blossoms, in settler and neoliberal imaginaries, are objects that imbue the landscape with color, beauty, and meaning in a land that was perceived as empty. These efforts call to mind the ancient Greek philosopher Plato and his concept of the ideal that is perfect and eternal, in contrast to what is present, emplaced, and real. Their beauty, as stated earlier, can be interpreted as a force of erasure for diverse beings (Carr and Milstein 2021), and yet also of genuine experience, in a meshwork of complexities that are human-vegetal relations. Trees, as part of the living world, become mirrors and agents of lived experience in place, and thus the influenza epidemic of 1918, World Wars I and II, the Depression of the 1930s, and the social media influencer all appear in human-jacaranda relations in newspapers across the years. Inhabitants of Grafton and other settler towns identify themselves and their meaningful experiences with the blooming of these trees in such a way that the blooming becomes entangled with humans in placemaking, both being relative newcomers. The color, scent, shapes, and atmospheres created by the trees in their blooming period shape human experience—the trees are affective in this regard. Jacarandas shimmer across the decades and shape human experience, culture, place, and sense making.

NOTE

1. While space and place are complex concepts, this chapter works with space as a more abstract concept, while place refers to a position that may have cultural and/ or subjective meanings (see Casey 1993). A fuller distinction is drawn later in this chapter.

Chapter 3

Wapato and Camas

Plant Influences on Wapato Island

I root around on this large river island for a celebrity, an aquatic plant called by many names: Wapato, Katniss (eponym for *The Hunger Games* hero), Arrowhead, Duck Potato, *Sagittaria latifolia, wakxa't* by the Chinook Indians, and many more. Having multiple plant names matches this species' diverse histories of both loss and plenty. "Wapato" was this plant's name among Chinook-speaking inhabitants along the Columbia River. Arrow-shaped leaves point skyward on thin stems with small white flowers rising above black waters with edible tubers underground. This plant's form is iconic, almost cartoonish in appearance. A plant with attitude gracing the marshy lands of Sauvie Island in Oregon, though wapato is native across wetlands in North and parts of Central America.

Wetlands on this island also root around, leaving traces of mineral-rich sediments and watery liminality through drier soils, such that you are often between worlds, breathing plants' humid transpiration as well as the larger system of evapotranspiration that is how trees and plants foster life as actors in climate and atmosphere. With the bounteous waters of the Columbia River and the minimal gradient, large floodplains are created with lakes, ponds, and sloughs. Sauvie Island used to have 79 lakes before white people arrived, and since then many have been drained for farmland. The shorelines of Sauvie Island experience daily tidal fluctuations and salt water intrusions, even though the Pacific Ocean is about 90 miles away. On this hot August afternoon, with snakes audibly moving in the grasses nearby, wapato's green vibrancy and tall stems rising above water speak to me, as colors do, of fecundity and generosity that can emerge from these metabolically rich wetlands.

In this chapter, I partner with the plant wapato, and to a lesser extent with camas, to delve into how they have influenced place and relationships in the Pacific Northwest region of North America. I draw on the phyto-human cultures co-generated with wapato and camas and Indigenous cultures in the region, and on two fields of inquiry that assess how plants influence humans—biogeography and psychogeography. As this volume argues, plant life aids humans in orienting to place and context, situating one in place and time, and also influencing animal relationships. This chapter is less analytical and more historical and descriptive, with some interpretation of my own experiences. Sauvie Island is a place I love to walk, visiting remnants of oak savannah and wetland plants; the island holds histories of phyto-human socialities that have been lost yet are actively being restored by local Indigenous people.

This chapter engages with naturalist and explorer Alexander von Humboldt's thinking and experiences in his plant explorations, and his holistic approach to the vegetal that focused on contextual factors as a way to know plants. The field of biogeography emerged from Humboldt's writings, which drew together aspects of geology and climate and more to understand why plants grow where they do. Humboldt asserted that each plant participates in sustaining its system, through adaptation to place and context. This perspective, taken up in the 1970s by James Lovelock and Lynn Margulis in the Gaia Hypothesis, is helpful as an explanation for how humans are distinct from other species—industrial humans are the only species not supporting their living system. In socio-ecological systems without industrial humans, the logic of the whole is apparent in each lifeway of each being. Wapato is a prime actor in supporting the whole and human inhabitants.

In 1805, when Lewis and Clark arrived here at this watery confluence of two gigantic rivers (now called the Columbia and Willamette Rivers), they named this "Wapato Island," as the plant grew in great abundance. In fact, this landscape hosted a dense human population that was only possible through the particularities of abundant vegetation and fish species in this floodplain. Indigenous land management techniques also supported the plentitude of wapato and another culturally significant plant with an edible tuber and a delicate purple bloom, camas.

Now these plains are not managed for wapato and camas, but are privately-owned, overpriced farmland. Lots of berry farms, fruits, and all manner of vegetables grow on the island. Oak savannah is in remnants here, a gentle and abundant primary local ecosystem that occupied 65% of the Willamette Valley, but is now 97% erased. In this era, when, as French philosopher Bruno Latour (2018) says, we can't access the Terrestrial or we cannot find a way to belong to the Earth, perspectives on plants offer possibilities to join the world with tendrils of knowing the community that makes up one's place (figure 3.1).

On a humid June afternoon, sitting near a hundreds-year-old Oregon white oak, I feel the sunlight and heat, as no other trees compete for sunlight here.

Oaks like to bask in sunlight. Tall grasses are beautiful and itchy on my legs. This oak is fairytale-like, with giant branches twirling, spiraling off the trunk in a bulky mass so that gravity appears to be defied—an engineering miracle. Their craggy shapes are a sight, and I observe for an hour or so, watching nut-hatches and cedar waxwings come and go. Being one of the last of what was

Arrowhead.   Sagittaria latifolia.

Figure 3.1   Wapato, from *Field Book of American Wildflowers*, 1909. New York: Putnam.

Figure 3.2   Oak Tree on Sauvie Island. Photo by Elizabeth Oriel.

once an abundant ecosystem that supported humans and other animals with acorns and other foods is a tense and emotionally-charged position, and I feel this welling in my throat. I have a field guide to oak tree systems in California that presents plants, animals, and fungi associations in story and science. Oak systems are diverse and plentiful, and I see how these two qualities support each other (figure 3.2).

Next, moving along, I perch on a wetland's edge, viewing wapatos' sensuous forms. Their body language, with tall stems emerging from water, suggests possibilities of unique grace in the muck of marshlands, of strength and growth. The arrow pointing up feels optimistic, affirmative. These traits track with their biological tendency to produce many more offspring than will end up reproducing, a perhaps intentional cornucopia for ducks, swans, muskrats, and beaver. They certainly epitomize plants' generative qualities as placemakers by providing for others. Ducks and swans enhance wapato's overall strength and resilience through their nutrient-dense feces, improving soil fertility and structure and controlling pests and weeds. Ducks also stir up mud in the water, which provides a more even distribution of nutrients. In China and other countries, ducks are raised in rice or paddy fields, as they have numerous beneficial effects on aquatic plants (Jiaen et al. 2017). Floodplain lifeways intermingle, collaborating to enhance strength and vitality.

My exploration on Sauvie Island is not a botanical hunt like Carl Linnaeus made in 1732 in Sweden, leading to his hierarchical taxonomies, with names reflecting genus and species. Nor is my time like Lewis and Clark's encounters with wapato and Indigenous populations that they called "Wapato Indians." I am not here to categorize plants or to map and lay claim to others' terrain. Rather, inspired by Humboldt's ethos, I am here to explore plants as subjects, asking how these vegetal beings influence and shape people and place.

It is all too common in the Western academic canon to claim that knowledge of the vast complexities of the world was attained by white, European scientists and philosophers. The sphere of knowledge contained in the field of biogeography, which helps explain why plants and animals live in a particular place or geography, was certainly intimately known to Indigenous peoples the world over. For those living in mountainous areas, the change of plants and animals as one moves up in elevation would be well understood, and the same for those living in coastal areas, for life on the edge of waters, in the shallows and in the deeper depths (Lomolino 2020). Humboldt lived among Indigenous people and learned from them during his time in Latin America, and yet he received earlier influences as well, such as in his friendship with Johann Wolfgang von Goethe. The complexities of the world could be explained by geography, Humboldt explained in his prodigious written works. Plants, especially, revealed the complexities of geography, of climate, topography, soils, volcanic histories, and thus help to orient one to what makes up a place.

Those who practice permaculture, which re-articulates Indigenous knowledge, can tell about the geology and landscape histories based on what plants they find. At least that has been the case before non-native plants exploded in numbers, altering plant communities. Plants have roles in enhancing soils, rain, moisture, and animal life; in degraded places, certain species such as dandelions loosen soils while wapato removes toxins. Plants also have affinities with other species, forming communities of shared resources. Scientist Suzanne Simard studied the sharing of carbon between Douglas fir and alder trees in British Columbia. Plants also have preferences in their growing and interactions, such as oak trees liking sunlight and growing less well in close proximity to others. One can read a place or at least make educated guesses by recognizing the plant species present.

Another angle on this theme of vegetal roles and influence is from psycho-geography, which is only very recently beginning to foreground plant life. Psychogeography interacts well with Humboldt's biogeography, as both see multi-directional lines of influence that include nonhumans. Psychogeography has explored influences of constructed space, in architecture and urban design, on human psyches and emerged from avant-garde circles in the 1950s and 1960s. The founders attempted to experientially access the autonomies of places, upsetting capitalism's hold on geography. Artists wandered in cities, drawn to sites where experiences are heightened. In work on cacti and their worlds across humans (Margulies 2023), to landscapes (Cooper and Kruglikova 2022), psychogeography has opened up nonhumans and rural places more fully. These two lenses open up a storied island's past in the United States' Pacific Northwest.

SAUVIE ISLAND AND WAPATO FLOODPLAIN ECOCULTURES

Curious about this island's past, I hope one day to see Indigenous stone carvings and to find certain traces of the British empire here. With Hudson's Bay Company's arrival here, the British were the first to lay claim to this island and constructed Fort Vancouver in 1825 across the Columbia River, which became a locus of trade moving from Britain to New York and across the Pacific. The British chose for their fort a location that had been a camas growing and gathering site. On Sauvie Island, British-run dairy farms produced butter to sell to Russian fur traders in Alaska, and French-Canadian Laurent Sauvie, eponym of this island, worked as a cowherd here and lived with his Indigenous wife, Josephte.

This island had been home for millennia to Chinook-speaking people in five villages, and the surrounding region was heavily populated. A Hudson's Bay Company official remarked,

> The population on the banks of the Columbia River is much greater than in any part of North America that I have visited as from the upper Lake to the Coast, it may be said that the shores are actually lined with Indian Lodges. (Simpson 1931:94)

Sauvie Island and the location of Fort Vancouver were central to Indigenous movements in the region, with many traveling downstream to access wapato and camas, and with movements upstream to fish for salmon. The Indigenous inhabitants on what is now called "Sauvie Island" and in this region were unique in that they lived quite sedentary lives because plants and fish were abundant, meaning they had no need for seasonal migrations.

Sedentism and living in towns were thought to be a result of irrigation and state control, though new evidence suggests that sedentism took place 4,000 years before state creation and resulted from living near wetlands with abundant plant life (Scott 2020). Dwellers in marsh areas, he says, have been considered primitive by advancing empires, which remove human settlements, drain marshes, and create grain agriculture that has been considered the foundation of "civilization." Yet, floodplains and their local abundance have offered human inhabitants self-sufficiency. In Scott's description, floodplains are political spaces in a distributed assemblage of actors and forces. This correlates to findings presented in chapter 1 about the percentages of forest cover and edge effects that allow Asian elephants their autonomy in Sri Lanka. Ecosystems foster local autonomies. Thus, the living world is a space of subjectivity, social worlds, as well as on-the-ground physicality.

Wapato and camas are two primary plants that allowed for a sedentary lifestyle, thus influencing and orchestrating social relations across humans and place. Wapato and camas are geophytes, as are potatoes, onions, and ginger. These plants store energy and water underground in tubers, allowing them to live in arid or alpine environments. They can survive through difficult spells of drought or bad weather. In this way, they are tailored for risk and survival, which allows them to become an asset and gift to those who rely on them for food. They often can be stored for long periods and provide carbohydrates during cold period, inclement periods. These characteristics are another example of plants providing for humans through hard times (as in chapter 5 on ash trees), and thus offering aspects of thriving and autonomy.

Humans need carbohydrates to live as they are a primary fuel source that human bodies require. As a tuber that tastes like potato or chestnut, stores well, and is easy to bake, wapato had and has been a *cultural keystone species* (Garibaldi and Turner 2004). For millennia, this member of the water plantain family (Alismataceae) has been a significant actor in diverse lifeways and uses in economies, rituals, and culinary habits for Indigenous inhabitants in the region.

Wapato is also an ecological floodplain actor, securing sediments and soils in shallow zones, which fosters abundant life. Globally, floodplains have shrunk down as rivers are dammed, with diversions and wetland draining. And yet, flood-prone plains are political and cultural shapers of human social organization. Ecologically, they work a kind of magic in balancing extremes of flood and drought, allowing groundwater to build up gradually, and plant life to flourish. Water movements slow down in floodplains and in this slowing, water interacts with plants, animals, and insects with less destruction. Socially and culturally, floodplains have fostered sedentary life for humans.

Cultural keystone species are akin to ecological keystone species, yet are situated in an ethnosphere, which is defined by Davis (2001:8) as ". . .

the sum total of all thoughts, beliefs, myths, and institutions made manifest today by the myriad cultures of the world." Though the prefix "ethno" refers to an anthropocentric construct, a multi-directional influence and agential lens approaches the ethnosphere as emerging with and from the biosphere, such that plants, water bodies, and humans co-construct culture. The social/cultural and the ecological are not distinct realms but are interwoven. For example, the Katzie people who lived further north in what is now British Columbia, learned of wapato's edible tubers from sandhill cranes, by observing the cranes digging them up (Lyons et al. 2018). Human culture learns from other animals, thus revealing one small example of a lack of distinct boundaries around human culture and other beings.

Wapato and camas as cultural actors were hugely abundant due to Indigenous land use practices. The physical appearance and presence of both plants attest to their cultural influence and qualities. This is described by Paul Kane, who visited Fort Vancouver and wrote:

> the only vegetables in use are the camas and wappatoo. The camas is a bulbous root, much resembling the onion in outward appearance, but is more like the potato when cooked, and is very good eating . . . They are found in immense quantities in the plains in the vicinity of Fort Vancouver, and in the spring of the year present a most curious and beautiful appearance, the whole surface presenting an uninterrupted sheet of bright ultra-marine blue, from the innumerable blossoms of these plants. (Kane 1859:186)

The Lewis and Clark journals mention that local people lived on fish and wapato, and that wapato was the principal article of trade. "The natives of the seacoast and lower part of this river will dispose of their most valuable articles to obtain this root" (wapato) (Lewis in Thwaites 1969 4:222).

Wapato is even central to certain tribes' origin stories, as with the Katzie, mentioned above (Lyons et al. 2018). Fifty km upriver from Vancouver, BC, a wapato garden was created 3,800 years ago, which the Katzie maintained for 700 years. Their origin story explains that their association with wapato came about from the marriage of two sandhill crane sisters with their cultural hero, Swaneset. This is an example of a relational origin story, across wapato and sandhill cranes who maintain close ties. These stories set the cultural tone for how one relates to the living world.

Archeologists Lyons and et al. (2018) profess a love that the Katzie people possessed for wapato, describing their cultivation of wapato as a love story, with intimacy and love across humans and plants. This feminist approach to natureculture relations foregrounds love as a primary mover of individual and collective action and takes an approach that incorporates mind, heart, body, and spirit. The relations between wapato and Indigenous people who harvested and relied on the plant for food do seem to carry layers of intimacy in close attention and respect.

The social structure of Katzie society was embedded with wapato (Lyons et al. 2018). Women were largely in charge of caring for and harvesting the plants, and certain families were in charge of wapato trading. After the British arrived, land was drained for agriculture, and the culturally significant plant from Europe, the potato, supplanted wapato. This change from the native floodplain plant to the non-native tuber altered land base relations as well as family structures and social cohesion (Garibaldi and Turner 2004). Wapato had a significant role and influence in structuring human social gatherings, community processes, and relations in these regions. The potato, native to the Peruvian and Bolivian Andes, was a primary cultural food for the Incas 1,800 years ago, and was brought back to Europe by Spanish invaders. Reliance on the domesticated potato led to the blight and famine in Ireland in the 1840s, a complex story that also involves colonial rule and oppression. This is the opposite kind of relationship found with the wild, yet cultivated wapato, which brought the possibility of sedentism and autonomy to the local people instead of dependence that would lead to many human deaths. In the Pacific Northwest, supplanting wapato in human diets changed the whole relation to wetlands as primary sites for nourishment and culture.

This role wapato and the ecotone of watery places play in Katzie social structures and lives is akin to the role newly growing grass plays for Asian elephants in Sri Lanka (more on this in chapter 1). When water levels recede from the large and ancient tanks or water bodies built in the dry zone, new grasses grow on the tanks' slopes. This, for many years, has drawn several hundred elephants from the entire northern region to a gathering point, in an annual social event. Elephants favor new fresh grasses and thus the grass draws them in, yet what ensues is remarkable for the social interactions of greeting, play, and bonding. Wapato has similar social relationships with humans, drawing them and structuring socialities.

Wapato was harvested two times a year, in both spring and fall in the Lower Columbia region around Sauvie Island. Not all Indigenous tribes viewed wapato as food. The Upper Nlaka'pmx people near Merritt, BC, used wapato in love charms and witchcraft, while the Lower Nlaka'pmx ate wapato as food (Turner et al. 1990). Upper Nlaka'pmx elder Annie York explained that wapato was related to other wetland plants, such as water knotweed (*Polygonum ambibium L*), yellow pond lily (*Nuphar lutea*) and marsh marigold (*Caltha leptosephala*) (Turner et al 1990).

Wapato and the great abundance of this plant's context, in alluvial soils with volcanic elements, and climate, along with human efforts to maintain large wapato stands, have shaped the culture and the place for millennia. Wapato was one of the central means by which the local inhabitants gained wealth through trading with those upriver. It is no wonder Lewis and Clark named this "Wapato Island." In fact, plants have often informed place names and also form landmarks for navigation. In Ireland, among 16,000 place names, 13,000 are named for trees (Living Tree Educational Foundation, n.d.). Before settler presence, Sauvie Island's history is largely shaped by wapato, yet, today that influence is barely visible or accessible. Wapato remains but is invisible to the local dominant settler culture.

CHANGING PHYTO-HUMAN ECOCULTURES

Beginning in the late eighteenth century, contact with Europeans and colonists arriving in the region brought several diseases—malaria, dysentery, smallpox, and measles. Malaria, in particular, brought such death to Indigenous people that by 1836, Sauvie Island was uninhabited (Hajda 1984). Between 1805 and 1840, 86% of the native population was lost (Darby 2005:13), and while in the 1820s Indigenous inhabitants in the region were in the majority, epidemics meant that by 1840, they were in a small minority. Signage I encounter along the Willamette River at times mentions Indigenous inhabitants before the British and later American capture of this area, though Indigenous removal to reservations and the cause of these diseases from settlers are not mentioned. Accurate histories can be difficult to access.

The arc for wapato followed the arc of local Indigenous people, and with the loss of their cultural traditions from colonial degradation of local cultures, wapato lost ground to the potato as the primary carbohydrate. Robert Brown visited the Pacific Northwest in 1865 and wrote:

> The roots of the Sagittaria sagittaria, Linn., were at one time very extensively eaten by the Indians, under the name of Wappatoo; and on the Columbia River there is an Island called Wappatoo Island, from the abundance of this plant. Since the introduction of the potato the use of the roots of the Sagittaria has much declined, and the name is now transferred to the potato. In the vicinity of nearly every Indian village are small patches of potatoes; but the ground is merely scratched up, and the cultivation far from being properly attended to. (Brown 1868:379)

He states that the potato is the only locally cultivated plant. The story of the potato as a place maker is quite a different one from wapato, being tied so closely to colonial diets and colonial practices. Irish artist Deidre O'Mahony created an exhibit of the ecocultural relations of the Irish with the potato, including aspects of shame that coincide with histories of colonial rule and the injustices of the imperially imposed famine. Phyto-human relations in situated locales are spaces of poiesis, and art like O'Mahony's speaks to these shared realms.

These phyto-human realms can be accessed with biogeography as a lens, in which the geologic and climatic features of a place take center stage. Sauvie Island's fertile, alluvial plains, just north of Portland, Oregon, are unique in that a large river, the Columbia, traverses an active volcanic range in the Cascade mountains, with Mount Hood, Mount St Helens, and Adams Peak all visible from this region. In the way that rivers move huge amounts of sediments, creating landforms from the force of water, the Columbia River carries sediments downstream that have built up island and riverbank landmasses, as has volcanic activity. Sauvie Island is part of what is called the Portland Basin, which is an alluvial landmass that has also been shaped by the Missoula Floods, 15,000–20,000 years ago (O'Connor et al. 2020). The dams on the Columbia River created in the last 100 years block sediment flowing downstream, ceasing the process of building back what water washes away. Thus, Sauvie Island's edges are eroding in places.

Wapato went from being a mainstay of Indigenous inhabitants' diets and a central player in trade across the region to becoming almost rare with British and American arrivals. The deaths from malaria, which decimated local cultures, also led to British practices and culture taking a larger place in socio-ecologies. The colonizers' practices of remaking the captured territories in the style of their home were evident here in potato cultivation. Another factor in wapato's decline was that Indigenous people who survived disease outbreaks were removed from their land and sent to reservations. Finally, land development and wetland drainage are factors. All of these changed the human-wapato relationship and altered wetland environments.

Wapato is not consumed to any large extent today. The plant is considered by both ethnobotanists and Indigenous people in the region to be important for ecosystem restoration (Garibaldi and Turner 2004). The Columbia River is heavily polluted, as is the Willamette, with erosion and loss of shorelines taking place. Wapato, as an aquatic and semi-aquatic plant, helps to stabilize sediments that flow downstream, helps to balance moss and phytoplankton levels in the water, and removes toxins such as heavy metals from soils. The plants provide food and shelter for insects, fish, amphibians, and other aquatic beings. Thus, wapato is a world maker for ecocultural floodplain lifeways.

HUMBOLDT AND WAPATO

Alexander von Humboldt never traveled to what is now Oregon, though he visited Thomas Jefferson in the United States in the spring of 1804, on his way back to Europe from a five-year-long journey through Latin America.

Humboldt's writings on plants in his volume, *Views of Nature,* and especially in the chapter, "Ideas for a Physiognomy of Plants," are a model for a relational approach to plants. Plants are both subjects of scientific inquiry as well as agential beings who influence place. He was a founder of the field of ecology, and *Views of Nature* has been called the first ecology book. His writings approach plants in an intersectional manner, such that ecology, politics, climate, and land use intermingle as they do in the real world. Humboldt eagerly visited the United States, yet did not approve of slave-owning and also critiqued the new country for the extent of extraction on the land. In Humboldt's worldview, nature and politics were not separate realms, as is also the case with contemporary scholar James Scott (2020), who writes of the politics of floodplains.

On his travels in Latin America, Humboldt was deeply affected by mining operations, which rested on the labor and the backs of Indigenous slaves. He had graduated from the Freiberg School of Mines in 1792 and then worked as a mine inspector for the Prussian government (Wulf 2015). His concern for men laboring in mines led to the creation of a free mining college, paid for with his own funds. Later, having met the Venezuelan Simon Bolivar in Paris after his five years abroad, Humboldt discussed and somewhat encouraged Bolivar to amass an army to liberate the land from Spanish imperial control. Though he expressed concern about whether he could find enough solidarity among diverse local people who had been divided and splintered by colonial rule (Wulf 2015).

It is interesting that this man, who brought plant life to the fore in all the forces and complexities at work on the earth, was against imperialism and slavery. He valued ground-up approaches that support underclasses and learning from diverse perspectives. One could say he had an ecological approach to learning and living. He seemed to be learning from plants, not about them, which is what led to his success in accessing and presenting a paradigm-shifting knowledge set. So many place names in the United States are named for him; even in the Pacific Ocean, the Humboldt Current runs along South America. He was a force in framings and knowledge production about living systems.

Plants are the dominant life form by far. Plant life makes up 82% of the Earth's biomass. This abundance has been central to lifeways on and around Sauvie Island. Native people there lived well off the alluvial plains where wapato grew, likely with a density of 28 plants per square meter (Darby 1996). Plants, Alexander von Humboldt argues, organize into specific forms, or appearances, that dictate the form of Nature. As he explains:

> The physiognomy of Nature is determined primarily by sixteen plant forms.
> I am enumerating only those that I observed on my travels through both

continents and over the course of years of attention to the vegetation of the various areas between the 60th degree of northern latitude and the 12th degree of southern. (2016:150)

After he studied plants throughout parts of America and Europe (1799–1804), he observed that their appearance reveals deep earthly processes, and that they speak, having an influence on animals and place. Plants are foundational for Humboldt, as he expresses:

> Plants ceaselessly organize the raw material of the Earth, preparing to mix together, through the force of life, that which after a thousand transformations is ennobled into animate nerve fibers. The same scrutiny that we devote to the spreading cover of plant life reveals for us the fullness of the animal life that is preserved and nourished by it. (Humboldt 2016:140)

In other words, plants are agents of organization of earthly materials and also help to organize and nourish animal lives. He begins a chapter on plants by describing the sheer abundance of life on Earth.

> When a person possessed of an active mind explores Nature, or ponders in imagination the broad range of organic creation, no single one among the manifold impressions that occur to him has so deep and powerful an effect as that of the ubiquitous abundance of life. (Humboldt 2016:135)

And plants dictate the mood or emotions of a place in Nature and how one feels in that place. In this way, cultural practices and forms, as in poems and songs, result from this influence of nature and plants on humans. Humboldt explains:

> The poetic works of the Greeks and the rougher songs of the old Nordic tribes owe much of their individual character to the forms of the plants and animals, to the mountains and valleys that surrounded the poets, and to the airs that swirled around them. Who does not feel a different mood in the dark shade of a beech tree, upon hills crowned with lonely firs, or in the middle of a grassy meadow, where the wind rustles in the trembling leaves of a birch? These native forms of plant life call forth within us images that are melancholy, solemnly uplifting, or merry. The influence of the physical world upon the moral, the mysterious interworking of the sensory and the extrasensory, bestows upon the study of Nature, when lifted to higher considerations, a charm that belongs to it alone, and that remains too little acknowledged. (2016:148)

In this last excerpt, Humboldt's biogeography intersects with psychogeography. The contextual earthly processes from which plants find habitability to spring forth and the moods and emotions those plants create are interrelated.

Speaking of the mood one feels under a beech or under a fir tree reflects psychogeography related to plants and their appearances and qualities woven with their surroundings. He calls on both, and in so doing, unites the physical and social sciences with art. In Denmark, where I live, the forest near my house where my neighbors walk every day is actually a tree plantation, with monoculture stands of different species. And yet, in this manicured and ordered place, the oak stands have a noticeably different quality than the beech stands or the silver birch or Norway maple that can be felt even in winter. These differences in atmospheres that trees project are not only from leaves but are from trunks, branches, colors, scents, light and wind interactions, shadows, leaf litter, mosses, and lichens.

Appearances of plants, that Humboldt emphasizes, are philosophically important in that how a plant appears to a person dictates how the plants will be treated. This appearance comes about in forms and perceptions of forms but also in values and stories connected to these appearances. For example, in chapter 5 on ash trees, university students in the United Kingdom are reported to not think of trees as beings that would have a disease. In other words, their appearances are not thought of as similar enough to humans to have illness. The loss of phyto-human cultures leads to plant appearances as alien, as simplistic, and as not relationally tied to humans (figure 3.3).

One of Humboldt's primary legacies and influences comes from his understanding of the vast complexity of nature and interdependence as a primary feature of living systems (Wulf 2015). This contrasts with reductionist approaches to plants, as when findings in botany are sectioned off from ecology, art, and cultural studies. He met another holistic thinker, Johann Wolfgang von Goethe, when he was young, and visited his home often as a friend and colleague. They shared a passion for plants and both studied patterns of vegetal life, searching out their forms to access what these forms mean and say. Both vegetal thinkers paved the way for Darwin's theories and his *Origin of Species*. Darwin carried Humboldt's books on the Beagle, providing his main intellectual framework, template, and inspiration (Wulf 2015).

Both Humboldt and Goethe found affinity with Spinoza's holistic vision of the world, and both men sought, in their approach to knowledge, an integration of intuitive, rational, and artistic approaches. Goethe's *Metamorphosis of Plants* (2009) looks for an internal archetype of form and development of each plant's life and finds two forces at work in plants: intensification and polarity. Intensification refers to the movement toward greater complexity, such as in development from cruder forms of stems and leaves to more elaborate flowers and reproductive systems and a greater refinement of sap. Polarity is a dynamic relation of opposites or a kind of pulsing between expansion and contraction that characterizes how plants develop with modes of moving outward and inward (Goethe 2009).

Figure 3.3   Monoculture Stand of Trees in Winter. Photo by Elizabeth Oriel.

What Goethe came to understand about plants has relevance to psycho-geography and to the impacts of place on animal emotions and psyches. Plants are not static forms but are growing, changing, metamorphosing, and this process has effects on those in their vicinity. A farmer friend told me that she knows what time of year it is based on the color of leaves. How is it possible that lines of influence run in only one direction, from the human to the earth? A linear line of influence and control from humans to the inert world dictates how dominant society treats the Earth, and yet lines of influence are always multi-directional. Drawing on French philosopher Jacques Ranciere, Jane Bennett (2010:vii) suggests that when we divide the world into dull matter and vibrant life, we are partitioned in our awareness and our sensibilities toward the more-than-human. Yet, if that lack of sensibility dissipated, what would humans experience of plant life? Goethe's scientific method includes imagination and intuition, combining form and function as ways to inhabit this broadened field of vitality.

Goethe writes of the "true Proteus" in vegetal life, a field of formative forces that guide the development of leaves, stem, flowers, and pollen. (The adjective "protean" means versatile, mutable, and adaptable, after the Greek prophetic sea God Proteus.) The parts of the plants are all united within the Protean archetype. Thus, each part is both unique and synonymous with the whole. Both Goethe and Humboldt approached plants and the living world from this perspective, tracking the relations between the parts and the whole. For Humboldt, it was between the plant and the geographical and contextual processes that help to explain their relative position, while for Goethe it was the internal forces and processes between plant parts in their growth and a unifying concept of form or archetype. Anthropologist Gregory Bateson, who is mentioned in relation to recursion (see Introduction), describes the sacred as the relation between part and whole or the "pattern that connects" (Bateson and Bateson 1987). Humboldt, Goethe, and Bateson were each occupied with this relationship of part to whole.

WAPATO AND PSYCHOGEOGRAPHY

Wapato and the marshy land where the plant thrives have an influence on my psyche. Wapato appeared as an actor in my dream one night—a stalk of the plant with an arrow-shaped leaf was housed in a wood and glass herbarium case, preserved and bereft of vitality but still visible in form. The dream indicates to me that vitality is a result of relationships between one's surroundings and one's own blueprint, the form of each person in their DNA. The plant speaks of context and lack of context in the glass case, and of domestication as a kind of glass house that many of us inhabit. Returning to visit wapato on Sauvie Island on a hot summer day is a reunion with a cultural icon. No one is around, and yet the plant exudes a strong presence here that reflects strong relations with context, in feeding birds, in holding banks together, in fostering wetland life.

I am drawn to this plant due to a sense of heightened experience in my recognition of wapato's appearance and presence, as psychogeographers have brought recognition to these types of experiences. Drawing on Baudelaire and his wanderers that drift through space, Guy Debord (1956) coined the term "psychogeography" in 1956. Over the years, this field addresses where psychology and geography meet, where place influences one's emotions and behaviors. Debord's work pushed against the alienation that capitalism inflicted on people and places, and his *Society of the Spectacle* is credited with inspiring student protests in Paris in 1968. Emotional connection to place, some argue, is the devalued glue or medium of attachments and of ethics. These connections have vanished in many places, leading to eco-logical crisis and destruction (McIntosh 2004; Smith 2016). In this way, psychogeography of rural places can be considered a kind of quiet activism, asserting one's emotional ties and attachments to the continuity of land's integrity. The very breadth of the relational and affective orientation to *place* that this volume takes up in regard to human-vegetal relations includes emo-tions, behaviors, thoughts, and all that goes into relationships with one's (or another's) land base.

Psychogeography has been posited as a means to help synergize scientific knowledge about the land with Indigenous knowledge (Cooper and Krug-likova 2022). This chapter speaks of this divide between white or settler land management styles and Indigenous ones pertaining to floodplains and wapato. The emotional impacts of place, and of nonhuman beings in place are part of Indigenous ontologies in which relations with land are central, and yet are not represented by other subfields in geography (Cooper and Kruglikova 2022).

Wapato's three-petaled, white flowers with yellow on their stamens are particularly charismatic. I stand on marshy ground and touch the slender leaves, feeling their strength and something vulnerable in their smooth tex-ture. Goethe wrote in *Metamorphosis of Plants,* that leaves of plants growing in marshy places have smoother and less refined leaves (Goethe 2009:19). Perhaps wapato's leaves are less refined, yet the arrow-shaped leaves point-ing skyward carry a refinement. Wapato removes metals from the soil and thus their leaves can contain metals, so it is not advisable to eat them, though testing on Sauvie Island reveals wapato is safe to consume here.

Something in this plant's physique calls to mind another beloved wetland plant, yerba mansa, a strong medicine, yet one that also removes toxins and even excess salt from wetland soils. My friend Hershel and I make pilgrim-ages to yerba mansa growing by the Rio Grande River in Albuquerque, New Mexico, and he is attempting to keep one alive on his windowsill. Shapes of leaves and favored habitats tend to call up other plants in other places, as biogeography does in tracing connections between similar plants growing in different locations and topographies.

These traces of relationships and memory are apparent in relational knowl-edge and memory landscapes, which Indigenous scholars write is central to their onto-epistemology. These correlations across landscape, memory, history, care, and emotion are a balm to modernist conceptions of the separateness of individuals as units. As Tyson Yunkaporta's book, *Sand Talk: How Indigenous Thinking Can Save the World* (2021) details how humans think, feel, and conceive of the world through navigating spatial orientations across relatedness. Wetland plants in the western United States share characteristics, in both physique and in qualities, and these qualities help shape human relations with them.

Another wetland plant and keystone species, the geophyte camas *(Camassia quamash),* has lived through a similar fate, from abundantly occupying terrain and feeding multitudes with intimate human-plant relations to a diminished and scattered presence. Camas are similarly delicate in form, with knee-high stems and bluish star-like blossoms in springtime. Leaves are narrow, like grass, and emerge from the base. Lewis and Clark wrote that seeing fields of blooming bluish-purple camas gave an appearance or illusion of lakes and water. Camas bulbs were another critical carbohydrate for Indigenous people, though moose, elk, and deer also feed on them. When cooked in a pit, the inulin turns to fructose, and the cooked bulb served as a sweetener for other foods (Stevens et al. 2001). Camas can live for 20 years and are also slow to produce seeds, taking several years. They occupy both wetlands and Oregon white oak woodlands (*Quercus garryana*); these oaks and camas have both lost out as Indigenous fire and other management practices generating these habitats are much less active in most regions of the Pacific Northwest.

I have a rather intimate personal plant story related to camas, which is more formally known as camassia. A member of the asparagus (*Asparagaceae*) family, other names for it are wild hyacinth and Indian hyacinth. One day in May of 2019, I went to Sauvie Island to see if I could find camas growing, as they are known to frequent oak savannah habitats and wet meadows. I had no idea where to look exactly and asked the camas plant to lead me. I started walking on trails, moving past ancient oak trees, and realized how little camas was visible. If camas did inhabit the island, it could be a challenge to locate. But eventually, as I walked underneath one particularly beautiful oak with craggy branches, I found several camas stalks and the distinct bluish flowers that look like starlight. It seemed like a miracle, and that I had somehow been directed by another's will. I imagine everyone has intimate plant stories, though they are not often told out loud.

Is this experience of asking camas to guide me and then finding camas in what seemed like an improbable experiment an example of psychogeography? On Debord's terms, perhaps. In line with Debord, this experiment, if taken seriously, that I was led by the camas plant, relied on a relational quality with the plant that is outside of capitalist and societal norms and behaviors. This experimental success can be explained perhaps in Gregory

Bateson's work on landscape as spaces of energy inputs and outputs but also of communication and mind. In most Indigenous worldviews, plants are communicative, are persons, or more than persons. In fact, in many tribes, plants are the ones who are in charge of healing, making decisions about which plant species are needed to cure a sick person and instructing the human healer (Gagliano 2018). In this enlarged sense of personhood, psychogeography opens up fields of relations across sentience in which others' moods and qualities shape experience across time and space.

Work in bioacoustics by Monica Gagliano (2018) and others found that plants do create sound and communicate, and that humans can listen to plants. She studied how several plant species in a laboratory setting communicate and found evidence of recognition of the others who were present and sounds emitted that changed behaviors in the other plants. Gagliano has been attempting to bridge Indigenous knowledge with Western science and reports listening to plants speak while working with shamans in South America and also listening to an oak tree in California. Anecdotally, oak trees are mentioned in various contexts as a common plant species that humans develop intimate rapport with.

Indigenous tribes, such as the S'Klallam in Washington state, are restoring cultural prairie lands, planting camas bulbs and other species (Thompson 2022). The tribe announced that they will not harvest camas until it grows so prolifically that the land appears as a lake, as Lewis and Clark noticed. Women traditionally would harvest camas bulbs, usually with a stick, inserting seeds as they did so in the soil to propagate and burning fields to maintain plant growth. Fire is central to maintaining prairies, and yet with homes scattered across the land, using fire as a management strategy is challenging. Less than 3% of Pacific Northwest lowland native prairies and oak woodlands remain. A member of the S'Klallam tribe describes bringing back camas and other prairie plants as climate and food resilience (Thompson 2022), as these offer better nutrition than Western diets. When the land is stewarded properly, these plants flourish (figure 3.4).

Tying the threads of wapato and camas, biogeography, and psychogeography back to this volume's themes, plant influences are experienced both subjectively and within social frameworks. Humboldt revealed who plants are, in their shapes, appearances, functions, and roles, as expressions of topographical, geological, climatic, and general biospheric contexts. These contextual expressions influence places, as psychogeography details. Wapato and camas, both as food and medicine, as well as ecological and social actors, have been central to human relations to land and floodplains. With dam removal beginning across the western United States, floodplains can flourish again, and wapato will come back gradually with Indigenous stewards' help, protecting banks and providing for diverse aquatic life and humans. Wapato

Figure 3.4   **Oak Trees and Camas in an Oak Savannah.** Photos by Elizabeth Oriel.

and camas provide historical insights into phyto-human cultures and societies in which plants participated, exerting influence and a politics of mutual thriving. Not all aspects of these societies provide ethical models. For example, the Chinook tribes kept slaves, as did other Native peoples in North America. Yet socio-ecological realities of abundance and autonomy (at least for some) in floodplains provide keys to ecological approaches to land use and culture. Works by artists like Deirdre O'Mahony in Ireland also highlight nonhuman agencies and relationships across land, emotions, and the past. More art of this kind builds phyto-human cultures as resilient networks of vitality.

Chapter 4

Social Worlds of Willow Basketry

Weaving the world, then, turns out to be a matter of "making culture."

(Tim Ingold, 2000:347)

Getting to know a plant takes commitment, time, exchange, and sensuous practices. Touch, taste, aurality, and sight are all involved. I come to these conclusions from living amidst a willow basket weaving community in Denmark.

From 2023 to 2024, I conducted planthropology (Myers 2017), experiencing and observing relations between weavers, weaving, and willow plants. Willow is the central vegetal being with qualities that have a role in drawing humans to the tasks of weaving such that it becomes central to their lives. Other crafts are not usually quite as embedded in relations with the lively, living lifeways of their materials.

I present basket weavers' own voices in this chapter, using a polyphonic style. I interviewed seven weavers and present informants' words in a continuous format, yet organized by theme, with the author/researcher's own ethnographic notes and analysis interspersed. Readers have access to experiences and intimacies that come to the fore. And this format tracks with the theme of this book, sociality among and across species' lines. (My own words are in italic font, and informants' words are in regular font.)

This style draws on Mikhail Bakhtin's (2014) literary work on dialogism and is also inspired by Nobel Laureate Svetlana Alexievitch's journalistic work. Polyphonic style has been called an ethical presentation of interviews, with empathy moving in multiple directions across researchers, informants, and readers (Lindbadh 2017). A hope is that willow's own voice may be accessed here as a social actor and subject.

In keeping with the themes of this volume, this chapter explores willows' influences, such as how willow organizes social lives and highlights context. In this weaving community, the body is a central mode of relationality. Basket weaving is a corporeal activity. Working with willow rods, flexing them by "asking" them to bend with your thumbs and arm muscles to do what you want them to, takes physical strength and sensitivity to willow's own body in relation to what the plant wants to do. The human body forms a close gestural association with willow in weaving that is a unique kind of knowing and physical rapport.

Weaving with willow is a form of poiesis, which means to bring something new into the world. This newness refers to either physical objects like baskets and also meanings. Poiesis can be thought of as a language of phyto-human worlds. The baskets each have stories connected to their histories with certain plants, knowledge, and relationships. As described in the Introduction, poiesis is both active and passive, and weaving is also both of these —one may be sensitive to the willow rods, and also actively working them into a shape.

Eva is one weaver who says that she becomes more sensitive from her weaving with and growing willow (see Figure 4.2). Basket weavers seem to have relationships with this plant that border on the shamanic, in the sense of viewing the plant as a being with a strong presence that each has been captured by to some extent.

My ethnographic work in a basket-making community in Denmark involved taking classes, going to more informal group meetings, interviewing weavers, and helping cut willow rods in winter. Basket weaving is difficult, and it takes time to develop a sense of willow's will and interests and a sensitivity in bending and flexing rods in ways that they tolerate without breaking. Classrooms are mostly perfectly silent as working with willow demands concentration. I found that time flew by, and deep concentration combined with the physical demandingness of the process left me feeling quite renewed. My job, which is heavily centered on computer and mental work, can leave me exhausted, and weaving has a different effect.

Willow species that work well for baskets are native to the Danish landscape with wet soils. A weaving tradition has persisted here for millennia. Denmark does not have its own style of baskets, as England, France, and Germany do, but the tradition has been maintained. One interviewee curated an exhibit with the last remaining commercial basket makers who make many per day. They report that no one is taking up the craft. Basket weaving is too demanding on the body, and baskets from Asia are now too cheap to compete with. In the community I joined, members do not rely on a sole income from selling baskets. For most, this is a second career and some teach as well as sell baskets.

An ecologically and culturally significant plant for both for land and water interfaces and for human and other animal cultures, willow (in the Salix genus) is a kind of quiet benefactor. Most weavers in Denmark grow their own willow, planting *Salix americana, Salix purpurea daphnoides, Salix viminalis*, among other varieties. Cuttings are put into the ground, about 25 cm deep, and planted in rows as this helps the rods to grow straight, with fewer lateral branches. In the first year, several rods can be coppiced by cutting close to ground level, and the second year, around 10 rods can be harvested, and the next year, around 30. The colors of leaves and rods change throughout the year and also change after cutting and drying. Each willow plant can live for decades.

Willow has gifts for those in proximity. Willow improves soil quality by removing toxins and is used in phytoremediation, and also dries out overly wet soils. I visited one community that planted willow as a filter for a human wastewater system. Black bears in the Great Lakes region of the United States scratch at willow bark and lick their claws for pain relief (Pers. commun. Ann Filemyr 2023), and this is how Indigenous people first discovered the medicinal qualities in willow bark for pain relief due to the presence of salicylic acid. Yet with so many benefits, both willow and the long-held traditions of basket weaving have been undervalued in many ways. Basket weaving has long been considered a low-status skill for which you would be paid poorly. Willow and her attendant crafts are cultural underdogs.

One author cites the genus name, *Salix*, as coming from the Celtic *sal*, meaning "near," and *lis* meaning "water," as willow is certainly a wetland plant. In another estimation, *Salix* derives from *salire*, meaning "to leap" in Latin, referring to willow's fast growth. "The willow will buy a horse before the oak will pay for a saddle" (Denham 1846:436).

The symbolic meaning of certain plant species often derives from their appearance and stature. Plants have a body language, just as animals do, or at least we animals read plants as if they were animals. Branches that droop often connote a downward affect, like a face turned down toward the ground. The weeping willow tree with drooping branches has been associated with grief and sorrow in England. Weeping willow trees were a symbol of death and mourning in Wales. In the Bible, willow has been a symbol of woe and grief, and in Psalm 137, the captive Jews in Babylon hung their harps on willow branches. In China, willow has been associated with immortality, and Chinese farmers would ask for rain from Dragon King Lung Wang in a ceremony wearing willow branch wreaths, as willows grow in wet places. Irish harps were made of willow, as the wood is known to have a soul that speaks in music. Willow also incites one to dance.

In Somerset, willow is said to follow one at night.

Ellum do grieve,
Oak he do hate,
Willow do walk
If you travel late. (Briggs 1978)

Every village used to have a basket maker, as baskets and willow products were essential to functioning on farms and in homes. Remnants of willow baskets have been found in an Iron Age site in the United Kingdom, which may be over 2000 years old. Willow baskets are unbreakable and were used for fish traps, as coracle boats with hides stretched over them, and for storage and transport. In France and other parts of Europe, the bark was removed by hand before weaving. In England, willow basketry became significant in the war efforts, in airborne pannier baskets, for shell casings, for officers' kits, and in World War II, domestic basketry was forbidden, as so many were needed in the war. When plastics became available in industry in the 1950s, many basketry workshops closed and the craft declined. One of the weavers I interviewed spent time in an old basket-making area in France where long-held traditions were still in continuous use. Willow is a cultural plant, co-generating phyto-human worlds across millennia.

Most weavers grow their own, and growing willow improves one's local ecosystem. While teaching a Masters course on sustainability last semester at Aarhus University, the students debated whether making one's own clothes can be considered a form of activism, or quiet activism. This term refers to small, everyday practices that can instigate social change (Warner and Inthorn 2022). One woman argued that her crocheting is activism, as it removes support for fast fashion, even on a very small-scale. Most basket weavers I met make functional objects, and these contribute to reducing consumerism as baskets provide practical functions. Weaving generates self-sufficiency on a local level that can be viewed as a response to and turning away from the consumerist and economic growth model of the world while improving local ecologies. Most weavers I met, though, do not cite this role as primary for them. The activism of willow basketry is a phyto-human process, directed by both.

WILLOW: KNOWING A PLANT INTIMATELY

Below are informants' own words presented in italic font.

> *Dorthe: Willow is strong in its nature. It is very, very strong when you work, and when it is dry and you have shaped your basket, it is also very, very strong.*

It is shapeable and ready to accept changes. If you weave it wrong, you can take it out. And you can do it again. And again. And it is not broken.

Dorte: It is nice to have in your hands. Some of it more than others. It's soft and flexible when it's fresh. It feels good to look at. On Anholt, there is desert willow growing naturally. I like that willow very much.

Eva: Willow is a material that provides you with so many different ways of working with it. You can work it fresh, you could work it half green, you can boil it and peel off the bark and you can dry it and soak it again. So you have a material that is manageable all year round. If you keep it in the right way. And when you walk in a willow bed, you can feel the freshness and the salicylin. And you can actually feel your head freshening, walking down the rows. And there's such a nice feel to cutting it and storing it and each process has its own qualities and then the whole color scale is absolutely amazing. So, you get all the spring colors—red, yellow, green, even black, as you've seen. And if you then have the patience to wait until it gets the skin of an old lady, then you can start working. If you do it too quickly, it will shrink and be loose. But if you leave it for a while, then you can work it half green, we call it. And that means that you keep the colors and they will stay longer.

Inge Lisa: Willow has a lot of possibilities. You can make a lot of different things. You can make things very safe and secure in a way. And you know how to do it. And it can be just good like this, every day. But, you can also do a lot of many different things with willow. And I also find that of course, it has a lot to do with the person who's weaving. And some people are more interested in artificial things. And some are able to do that because they are more interested in the techniques they have. I have been in a way more interested in techniques, because it has something to do with who I am. I didn't know or I didn't dare to do experiments. But I think it's the person. The limit is with the person, it's not with a willow. Willow can do a lot of things if the person is interested.

I can make a round basket, an oval basket, a square basket, flat baskets, I can make a lot of baskets. So now what is to learn? Ane has been making new designs. Some of her baskets are her own designs, and nobody else's. But the willow is following her in a way. And I think that's very interesting. You can use the material in so many different ways. And there are many different kinds of willow. I think when we were growing it, we had about 50 different kinds, we really had 50. You can do some things with one kind of willow, but not with another kind. Each variety has its own characteristics. But willow has a lot of possibilities if the person is able to be creative?

Tim: Willow responds to how you treat it. One of Robin Wall Kimmerer's essays [2013] is about sweetgrass, and the way the Native Americans managed it. Different tribes have different ways of doing it, some cut it, some pull it out wholesale, or some were very selective in cutting it. However it was managed and grown, it seemed to respond by producing what the Native Americans valued in it. If you just left it, it didn't do so well. It reached an equilibrium and its natural life span then took over. So, the clump would grow and then get too crowded, too overgrown. And it would not thrive as much. And if they picked too much, then clearly it didn't like it either. There's that balance, that relationship. And I think willow is the same. If you look after it, if you coppice or pollard it responsibly, it will continue for many, many years to say thank you, to repay you, and you'll be able to use it. So, there's this commensal relationship between us and the plants. And I think willow is one of those plants that exemplifies this relationship. I like that aspect of it as a plant that has evolved for itself, but equally is appreciative of being looked after.

When I first started working with willow, I thought it is really a good thing to work with. It smells nice and has a nice feel about it. If you understand it, you can get it to do all sorts of complicated things that you wouldn't have thought possible. It's not imposing yourself on it. It's working with and respecting it. If you're arrogant with it, it brings you down to earth very quickly. Working with it is quite grounding in a way that working with cane isn't. Cane, it's machine made. If you work with whole cane that is different, you have to respect that. But if you're working with machine-cut cane, it has its uses, its properties that you can exploit, but it doesn't have the same soul, the same feel, it doesn't evoke the same response in me as working with willow. Growing my own willow made all the difference, and I find it difficult to put it into words, but it does make a difference. The fact that you watch that piece of willow sprout from a stump. You've gone and looked after it. Which was difficult when I was working, I still find myself very busy. So, I'm not on the willow patch as often as I would like to be. But then you select it, and you decide what you're going to make with it. And then you produce a basket. And something about that the basket is tangible from my point of view, adds something to it, which I try to convey when I'm selling the basket to somebody else, the fact that I've grown this willow. This is my willow. To know the person that's grown it and I can identify the different varieties.

Willow is similarly experienced across the weavers as being good to hold and touch, and also offering great variety in types and colors and in ways of working. The plant's qualities seem to be central to the relationships they hold with the craft, the skill, and the process. Willow's influences can be felt, as they describe the qualities willow brings to the process of weaving. My own pull to continue weaving with willow comes from an olfactory inspiration, in my response to willow's scent which is strong in a sweet and slightly dense way. The scents, which vary by *Salix* species are actually impossible to describe, but they remain with the basket for many months. As I come near one of my baskets, I feel an uplifting sensation from catching a whiff, which is delightful. An intimacy in this way moves from plant to human bodies, with olfactory influences.

In my own weaving, I develop a certain fondness for specific varieties, such as desert willow, which is blackish when dried with fuzzy buds that remind me of deer antlers. Though I like working with all varieties, some are more flexible, some longer and thinner. Each has its role in the baskets, as stakes, as weavers, or as parts of a thick and solid base. Traditionally, the base rods are called the grandparents, the stakes the parents, and the weavers the children.

Growing and weaving with willow is quite a demanding process. Willow needs to be harvested in winter before the leaves appear, when the sap is low and thus the rods are not rigid and are easy to cut. Most use a coppicing technique, cutting the stems just at the base with a knife or secateurs. The willow needs to be sorted into different lengths and dried for several months and then soaked before weaving, often for one to two weeks, or shorter times with hot water. This requires a large outdoor tub or plastic-lined hole in the ground. When making a basket, the willow rods need to stay moist. In warmer weather, we keep all the willow outside, wrapped in moist towels, and only bring rods inside for our next step. The demands of the material are specific and quite high, and bring each in relation to its outdoor context.

The plant is very present in all these activities and relations, redrawing boundaries of the social across both humans and plants. This harvesting, drying, soaking, and moving inside and out, I find, brings weavers into the atmosphere and lifeways of the plant, with great attention to willow's "perspective" and to the plants' bodily needs (or at least the needs of the willow in the context of weaving). The willow seems to remove the weaver from their indoor world and shifts the line of domestication. This process of civilizing, of drawing lines between those in and those outside, places the humans in the outer category for the duration of weaving.

Domestication has involved practices and narratives of drawing lines and boundaries, separating what is inside from what is outside. The term derives from medieval Latin and meant both to "tame," "to dwell in a house," and "belonging to a household." Archaeologist Gordon Childe wrote of the Neolithic Revolution as a time "when man ceased to be purely parasitic and, with the adoption of agriculture and stock-raising, became a creator emancipated from the whims of his environment" (Childe 1928:2). This revolution allowed for the "civilization" of Europe, Childe argues. This narrative has been foundational to colonial and postcolonial land use practices, breaking up situated and long-held relations with place and the land that fostered continuity in ecosystems.

Domestication and the erection of boundaries (Lien et al. 2018) can be seen as part of industrial humans separating themselves from what is "natural" and also justifying or allowing for a gradual shift for humans to identify with machines rather than with trees and the living world. In chapter 5 of this volume, an environmental educator describes how university-age students that she takes on walks to show ash tree decline have trouble thinking of a tree as something that would have a disease. Trees, for some young people, seem to be inert, barely alive, and not vital organisms that share similar susceptibilities as humans do. This is a kind of domesticated thinking, in which trees in a forest are outside of the laws and realms that humans inhabit. This breakage in awareness sits at the heart of the current eco-crises.

Weaving with willow troubles these barriers and brings humans into a plant's world. Social anthropologist Tim Ingold (2000) troubles the divide of artifacts or baskets from living things. Anthropologists and archaeologists tend to view artifacts as products of culture, and not as products of materials which have their own forms, ways, and approaches. Weaving and the products in the form of baskets are both autopoietic (Ingold 2000), such that the weaver engages with a system of self-organization that is dictated by the material, the weaver, the soil and minerals that the willow absorbs, the climate, and more. Weaving is indistinct from all forms of poiesis, or coming into being.

Becoming a basket weaver requires great commitment, including owning land or grow willow on a friend's land, and engaging in all the activities related to sorting, drying, soaking, and weaving. There is certainly a physical aspect to this boundary shift, yet there is also a subjective shift. These relationships have an aspect of intersubjectivity, which Edmund Husserl describes as an interchange of thoughts or feelings between two or more subjects that works through empathy (Cooper-White 2014) and also as an experiential sharing across subjects (Reuther 2014). While plants having thoughts and feelings may be a stretch, there are affective lines of attention and care across the human-plant divide and a sharing of experiences and meeting of wills between plant and human. Weavers are open to the material in an empathic way. Research on Indigenous relations with plants speaks of intersubjectivity between species. For example, Mentor (2012) writes of intersubjectivity between cassava and cultivators among the Waiwai in southern Guyana in their speaking of Cassava Mother, a being of mixed human-plant identity. This intersubjectivity was also apparent in the past, where I did fieldwork in Sri Lanka between humans and the rice plant (Van Daele 2008). Willow and weavers push back the veil of nature/culture separation and engage with plants on their own terms or on negotiated terms, as this is required to make a basket. This dissolving of boundaries is even more evident in the next section on communication across species.

This last informant speaks of cane, which comes from the bark of the rattan vine. These materials derive from tropical climbing plants in the Palm family (Palmae and Arecaceae). The largest rattan genus is Calamus, with 370 species, and comes from India, South China, and further southward, all the way to parts of Australia (FAO data). Many baskets sold in stores are made of rattan, and this group of weavers speaks of rattan basketry in less-than-positive terms. The material is processed and becomes completely uniform, having none of willow's unique characteristics.

PLANT COMMUNICATIONS

Ane: I was working in a folk school teaching psychology, and that was a very mental universe. I've always been doing creative things with my hands. But suddenly, I saw myself in this very mental universe, and just wanted to do something else. And then I had this dream that I should pursue the willow. In the dream, I was traveling from Denmark down to the Middle East. And then I met two rivers. And I was told by an old man that the names of the rivers meant "willow." And I had to go with the stream, with the willow stream. It's almost 30 years ago. I had never looked at a basket before. I didn't find them interesting. But the energy in the dream was so, so strong, that I committed myself to a weaving course. Immediately after a weekend, I found it and said, "I'm going there." And then it just took off from there. And I found out that it suited me so well. This more physical work with the willow, it's not like sewing. You have to use your whole body with your weight and that suited me very well, at that time. So, it took off, then it just went on and on and on.

Eva: I had a dream about making an International Basketry Center. And, within two weekends we turned an old pig style and horse stables into a workshop, and then started up and it was a great success from day one. I had two people helping to start the whole thing up. And so, the dream actually came true in the way that we built up the workshop. And in the years to come, I made friends with so many basket weavers from all over the world. I traveled a lot, and I invited them to come and teach. And after a couple of years, I invented the Big World Basketry Day where I invited people from all over the world, mostly from Europe, for a weekend. On Saturday, it was for the makers and the weavers and the artists to mingle. And then, Sunday was open to the public. And I did that for a number of years, and it became a great success with 1000s of people who came and shared their knowledge and insight.

Dorte: In the past, I would look at books about historical traditions in basket weaving, and was inspired by the designs. But now, the plant tells me what it can be. They speak to me through my hands. I just came back from Anholt and there were lots of roots blowing in a big storm. So, I walked down to the sea to find roots that were exposed. I made an installation from the roots, and that was very much what they wanted. This sort of willow (she shows me a basket she made), it is so easy to feel what the plant wants. The seaweed is inspiring, it is nice to have the material curving. The plants have a will.

Willow and other plant influences are visible here in what are expressed as direct human-plant communications. The first two excerpts describe having dreams that instructed them or directed them to work with willow in various ways, and the third feels the plants instruct her on how to make the basket or art piece. These influences are direct communications and are central to the artistic process. In each case, there is a trusting in these communications, and for each, the instruction or messages lead to remarkable results, such as careers as weavers and teachers, plus myriad art pieces formed through plant instruction. For one, phyto-communication led to a successful basketry center.

Several speak of the will of the plant, the process of becoming sensitive to that will, and learning how to collaborate with willow. The first excerpt is unique in how she received communication without any contact with the plant or with baskets. Gardeners and farmers could easily speak of plants' will, yet weavers access the plant's will informed through the body while weaving. The dreams and communications mentioned are examples of inter-subjectivity. An attention between species is apparent, attention to willow's affective qualities and influences, and even to willow's voice, if one listens and acts according to the plant's direction. These weavers speak of an inter-species collaborative effort. And these efforts become baskets in the world, become something new, a poiesis that fosters a poetics of cohabitation (Ribo 2022).

GETTING STARTED, BASKETRY WAVES, AND STATUS

Dorte: I started making things as a child. I've always been going to the forest and collecting plant materials and making installations at home. Anywhere I lived. I collect everything that I find interesting. I think it's because I'm from a farm. I am the last one of eight children and each one of us did something. My father was carving wood. Each sibling has had something that they did, some used coconuts. My mother drew a lot of pictures. We had a normal farm with pigs, cattle and grain.

Ane: So, it's a little sad story because if you couldn't do anything else, you could always be a basket maker. If you couldn't grow anything else on your land, you could always grow willow so it's kind of the lowest of the lowest position you could have and you will not be paid as well as other crafts. For example, a carpenter was paid much more than a basket maker. I don't know why it is but it is so in Denmark and in most the other countries from old times. It's never been really high social status work. I had a really good friend in England, an old basket maker. He became a basket maker because there was no place as a carpenter or electrician, the only apprenticeship he could get was as a basket maker. And he was paid half the wage as the others. And he came to Denmark and taught many times. And he knew all the different English baskets and he said, "I had to make so many baskets a day. I really envy you. You make one beautiful, artistic basket a day. I would have loved if that had been possible for me in my life."

Inge Lise: I thought it was interesting to do it. When I was still in school at 14 and 15 years old, we made baskets with rattan. At that age, I had to earn my own money, so I made baskets and sold them. And then when I left school I became a social worker quite quickly. And I worked as a social worker for many, many years. And then, in the 1990s, my husband and I went to a market where I saw a man doing basketry. And I said to him, I want to do that now. I looked at what he was doing and I talked to him, and I joined his course. And then a bit later, my husband also tried to do basketry. So, in a way, basketry became very important for us in the 1990s. I was still a social worker and a leader of a school for single women and it became too much to do both. So, I dropped my job as a social worker. I was something like 53 or 55 years old. And then I started doing basketry and growing willow. We had a place not here because you can't grow willow in the small garden here. We had a place north of Copenhagen, we could grow willow there on a friend's land. So, we were growing and selling and doing basketry.

Eva: I was a craft design and technology teacher originally. And at some point, I attended a course with Bent Winkler who is one of our great weavers in Denmark. He was working at these high schools, we call them homeschooler. I took a three-week course in willow basketry, and I was completely taken by it and changed my direction after that. There is rattan weaving, which is more for retired people to work with. There was no honor or quality in that. But the willow was new. We were taught how Bent had found willow and how he started growing it. And so, it became not only the basketry, but it became, you get your little cutting, put it in the ground, see it grow, cut it yourself, and weave your basket. So that became a whole new setup for housewives and others during that period of time in the early 1980s. And then 10 years later, there were about 800 weekly courses in Denmark, which was a huge number. It became immensely popular. I mean, there's always been waves of felting, knitting or crocheting or whatever, and it would take off. But in this period, it was definitely basket weaving. And it was, as I said, not only the weaving, but also the planting business.

This was the first wave of willow basket weaving, as Bent Winkler found wild willow growing and started making baskets and teaching. Before that, it had been a professional craft in Denmark though it was dying out.

Gitte: I was not born into academia, which is where I work. I grew up a working-class girl. I somehow take up things which were from former generations, to do craft that has been in my family. I apparently had the talent for academia when I was in school, and everybody could see it. I was encouraged. And other talents which were more common in my family were not encouraged. And I did, I enjoyed it. I liked that I could have success, but I needed to do other things. My grandfather was a bookbinder and others had been blacksmiths and farmers and my grandmother was cooking for people and things like that. These things, using your hands and producing something you can touch or eat, that's better. I also like cooking.

Ane: The old basket makers were not making much of themselves, and they didn't have a lot. They were not rich, of course they had their food, but they kind of used everything. So, I would say they belonged most to the poor end. And in my travels around the world, the old basket makers of my age, they all are humble. They are all humble in a way. I think it has something to do with the old craft and the willow, you go out and cut it, you can make something, and

Figure 4.1   Basket Weaving in Process. Photo by Ane Lyngsgaard.

you will never get rich on basketry. So, the people who choose this enjoy nature, enjoy the setup about making baskets and so it attracts a certain kind of person. Also, in the books and the Old English basket makers, they all have this humble attitude to life and to other people.

Eva: There was huge enthusiasm that had been around about willow basketry. And there were so many courses. And there are lots of moral and ethical questions that you could pose. The whole spirit was that if you've been to a course, you go home and teach it to somebody else. So, we were not doing four-year trainings, and it was not well respected abroad in Germany, or in France where they have a four-year school, and where they have very specific rules about what kind of basketry tradition you belong to. We have no rules. And if there were any, we broke them all. So, that was a whole explosion of this energy, enthusiasm. And I think that this attitude has had its bad sides, of course, because we also acknowledge that masters are good. And you can learn something from us and you can learn something from repeating your work absolutely. But we didn't think like that in Denmark. You went for a course and you made your weekend basket or two. And you were happy about it for the next year. And then you would seek another course with a different technique, and repeat the whole thing. The whole idea was so different from getting to learn and learn properly.

As I said at the beginning of this chapter, getting to know a plant takes commitment, and weaving with willow takes time, access to land, seasonal

Figure 4.2   Willow Growing in Winter, Denmark. Photo by Ane Lyngsgaard.

tasks, and more. In this way, willow basketry is quite different from knitting, as most knitters do not keep sheep in their yards. A number of the weavers grew up on a farm and see this as part of their weaving story; they learned the importance of making functional objects and they are well acquainted with physically demanding work that requires outdoor time. Many began with craft as a child, and their weaving is a kind of return. Most were already gardeners, so growing willow was not an entirely new skill to learn. Many draw a distinction between the physical weaving and the mental work of many jobs in society, which require long periods on a computer. And they are all aware of how basketry has been considered a low-status endeavor, though this doesn't bother them.

One mentions she has a working-class background, which she associates with basketry, inserting class consciousness into this human/plant social world. Forests have long been considered political spaces. Forests and plants have provided protection for rebels who resisted colonialism and were central to the ecofeminist Chipko movement in India, which sought to end deforestation. A common ethos of working-class life, especially those connected to the land, is self-sufficiency in making useful things. This aspect of self-sufficiency is another aspect of activism, in which they take back their lives from the requirements of the establishment and from more elite modes in society toward greater attention to the living world. Many speak of their basket-making inspiration coming from the look and feel of the finished product and also struggle a bit with making non-useful art objects.

This recalls Jane Bennett's words in *Vibrant Matter: A Political Ecology of Things* (2010) in which she writes that the image of matter as dead, as inert, feeds human hubris by "preventing us from detecting (seeing, hearing, smelling, tasting, feeling) a fuller range of the non-human powers circulating around and within human bodies" (p. ix). Weaving is a kind of eco-politics in how growing willow improves soils and local hydrologies. Though these weavers are not ecosystem people in that their livelihood is not coming from their land base, their engagement with historical traditions of phyto-human social worlds pushes against modernity's erasure of the past in favor of the growth possibilities of the future.

The explosion of interest in willow basketry in Denmark had its own unique qualities, with no real hierarchies of masters and pupils, but many novices teaching and spreading the skills. One interviewee says that this is typical of Denmark's approach to life that is ground up and non-hierarchical. This basketry wave opened opportunities for most of the weavers I interviewed, though most started later than the 1980s. This culture that grew up around a plant, and around growing and making with a plant, is a phyto-human culture in which both plant and humans collaborate. What is generally thought of as human culture is certainly a diverse multispecies affair.

As Michael Pollan argues, plants have domesticated humans as much as the other way around. The next section speaks of the mental and physical aspects of this phyto-human culture.

MAKING BASKETS AND PHYTO-HUMAN CULTURES

Dorte: That's funny, because before I started to make willow baskets, I always had white fingers. It was because there is not good circulation in my fingers. I have an autoimmune disease of the connective tissue. And when I started to use willow, I haven't had it anymore. My hands are fine now. Working with willow healed them. And I have had this since I was a child.

Tim: I like looking at historic baskets. Baskets are a very good example of evolution at work. They are highly evolved, because they have been designed to observe particular functions. And to exploit particular locally available materials. So, there's that story behind each basket.

Tim: When you're making a basket, it is total immersion. And you're using all your senses. But equally, you're not thinking about it very hard. Because when you're doing something, it is actually working at a subcortical level. A lot of feedback comes in from the basket, be it temperature, sound, and feel, what

Figure 4.3   One of Ane Lyngsgaard's organic baskets. Photo by Ane Lyngsgaard.

you're looking at. It's the biomechanical bits. How's it feeling under my fin-gertips? Is it bending easily? I've got to be careful of that particular bit of that particular rod, because I see that it's going to be difficult when I get to it. So, you've got to concentrate on what you're doing.

Ane: I think what I like is that you have to kind of go with the willow. And you have to focus on it. You're really present when you work with it, because you have to be present otherwise, the willow just takes its own way. So, it's not like you can knit for example, just sit and talk. I feel that you have to be present with the willow otherwise you won't get the basket you want. And that's a huge plus for many, many people, even if they are struggling, and it's hard. But they are so focused, that actually when a day is gone, they have just been in another universe. You might have tried the same with something else. You don't need a lot of tools, you don't need to saw, you actually just have the material, and then you can do something with it. Of course, until you're really good when you can do what you want. But I think the willow demands your presence. And that's good for our civilization, I think because we're so much in our heads. It's a gift for many people, even if they struggle, and it is hard work, you know!

The other thing is that being a basket maker for your whole life, if you make a living only from making baskets, it is really hard on your body. And I can see that on the old basket makers. Their shoulders, elbows, hands and back. It's really hard work. Also, the harvesting of the willow, the sorting of the willow. And as a teacher you lift a lot of kilos when you put it in water and get it out again. So, it's not by accident that the old basket makers were all men. The women made much finer stuff, seating or straw work or something. But the wil-low was actually for men.

When I make a basket, especially an organic one (see Figure 4.3), I have to feel that it is filled out from the inside, that it's not empty. So, the shape has to be formed from the inside. And that's different. It's a different thought than just making a shell. And that feeling or that thought has made me feel that every time when I'm really concentrated. I work on myself. How is my inner space? Is it filled? Or is it empty? Does it stand strong with the energy from inside? Or is it just a shell? And it's easy for me to feel the difference. If a basket is made with presence, when it's finished, it will always stand and shine. If it is made by your hands but not with presence, you can see that, and it can be a fine basket, but it will not shine. I have been much more aware of that. I can easily make a basket and not be present. But it will never get really beautiful. And you can say that knowledge made me think of being present in other situations? Because it's so easy just to do or talk without really being there. So that has been a process for me, I would say. I'm not a very patient person. But working with willow, you have to be more patient. Because if I am too quick, I would lose contact with it. It's about the rhythm in it and the repetition. It is about doing the same again and again and again.

And I will look at other trees and see, can I use that for something? Or can I use that? Or when I go to the beach, I wonder if I can use this seaweed in my weaving. So somehow it has opened my eyes to finding things to use, and I didn't do that before. To pick up a stone because it's interesting. But now I really, I never get out of this mode. I can walk for hours. I've also found out that you can use lots of materials, like nettles, straw, grass. So, my horizon has opened up for natural materials. And sometimes you just try something and see, does it work? No, it didn't work. So, you could say I'm more curious about plants. And also curious about what's living just close by me. You don't have to travel a lot. What grows here and what can I use? For example, I'm using my own lilies. And the rush in the field.

Living at a basket weaving school, I have had access to willow that is ready for weaving all day and night, with a freezer full of rods, as this keeps them moist after they've been soaked. I started a basket one night by going outside to the shed to find the right size and length sticks. I brought the willow to my room and started the base and a few inches of a technique called waling up the sides. But there are so many steps and little tricks, like cutting the willow's edges or bending the rods in a certain way, and I have trouble remembering them all. Learning basketry well enough to teach would take me years and years of practice. Strange that it is considered a lowly art compared to others.

Another aspect of the demands of willow is in how basket weaving requires such immersion or concentration, setting it apart from other crafts like knitting. The room is silent during a course, each working in full concentration. My fifth basket was a disaster, as the stakes lost their uprightness. I had lost the awareness of how each part fits into the whole. You need to be aware of both parts and the whole at the same time. Baskets can be molded in various ways, though the structure of the base, the stakes, and weavers must be solid and true. The social worlds across humans and willow are demanding in many ways, and yet the weavers seem to respond well to the demands and even appreciate them. The demands seem to reset the weavers' minds and bodies in certain ways.

Maurice Merleau-Ponty's intersubjectivity as a shared rhythm and as a shared "manner of handling the world" is present as weavers describe their baskets as requiring concentration with the rods, and that the negotiation with willow is like having a shared mind. The input of the mind and body to "working with willow" seems to produce a basket "that shines," one says. This quite intimate relationship requires sensitivity, empathy to the other, and inner strength.

An aspect of this intersubjectivity is in feedback loops taking place as one weaves, as one interviewee mentions, in how the rod feels on one's thumb and the constant negotiation with being flexible. Can I exert this much pressure on this rod? These feedback loops constitute a huge part of learning about the willow's will, what the willow will do and not do. And this is a recursive process, which was mentioned in the Introduction as a process that plants are also oriented toward, in Fibonacci sequences, for example. Recursion takes place through feedback loops that take place as lines of movement in learning between the weaver and the basket. The weaver is drawn back to their own past experiences, which inform the future experiences. Anthropologist Gregory Bateson described self-organization and learning as recursive processes, and this kind of learning is central to how traditional ecological knowledge forms over time. A response to the other is inherent.

Weaving also opens contextual relationships and new perspectives in relation to place, as all plants growing around the weavers are perceived as potential materials. I have seen baskets and other objects made from rush, larch, seaweeds, cork elm, roots from Africa, mosses and lichens, random sticks, dried flowers, and much more. This opening to the bounties of place in plant materials is present in each of them and draws the qualities and lines of place into a phyto-human social and artistic sphere. Each walk becomes part of their imaginative making process, one that finds its emergence between humans and plants. The next excerpts speak to the temporality of this social phyto-human world.

TIME AND WILLOW

Eva: I learned from growing willow, but mostly also from this friend of my family who have been basket weavers for seven generations. They taught me there was a whole circle of activities around the year. They grow willow, they harvest the willow, they sort the willow. In France, willow is peeled in June. So, you have a lot of festivities going on while that is happening, and then you weave it, etc. And then you sell it. And throughout the year there were a lot of other activities. And they were growing their own grapes. So, they made their own wine. And of course, they had specific baskets for wine grapes and collecting. They grew walnuts, and they would sit and play and sing while crushing the walnuts. And they would have specific baskets for that, which were made out of the peel of the white willow as well.

Ane: I go out in nature in times of the year where normally you wouldn't go out. For example, when you cut the willow, it's wintertime and it's really cold. And normally, you would just stay inside or go for a walk. But I'm working a whole day outside with the willow. I don't think I've ever done that. And so, this was a new experience. And it always feels good. Somehow you work when the light is there. And what you do is very physical and it has a purpose. I remember my childhood on a farm, you had to do this. And then you harvest it, it goes over a year. So, it has its own rhythm. So now it's time to harvest here, it's time for these and now it's time for that. It's a cycle.

Willow basketry has seasonal and cyclical roles and duties, drawing weavers into a circular relation to time. Some say they grew up with this cyclical time on farms. The necessity of cutting willow in winter is unique and brings people outside in the darkest days of the year, another physical demand that they all say they enjoy. A different rapport with winter is part of the relationship.

Phyto-human social worlds have their own temporality and heightened awareness of plants such that plants exert influences on contextual awareness. The capitalist, industrial approach to the world has its own time, organizing human bodies around work life and time off. This is a domestication of humans into units of time and inside lines drawn around what is civilized and what is not. Willow basketry undoes some of this domestication in relation to time, opening weavers to willow's own timings, to winter days, and to a cyclical, seasonal way of life.

Weaving is demanding of one's body and one's mind but in quite a different way from many jobs in the world. These demands can be hard on the body and require intense concentration, yet the weavers speak of the benefits and joys of this process. It seems to enliven them and not drain them. Time disappears when they weave, and they enter a kind of flow state, listening and responding to the willow's body and will.

Willow is a significant plant teacher, in the way that Robin Wall Kimmerer (2013) speaks of plants. This human-plant intimacy is an example of learning plant subjectivities. Listening and attending to plants' subjectivities is part of co-generating an ecological civilization, of inserting humans into the world. Learning with willow is a recursive process with feedback loops between the two bodies, human and plant. Taking cues from plants is a kind of attention to the Tree of Life in the Genesis story, and to avoiding the dualisms of human/nature and nature/culture. If land were organized around plant subjectivities, there would be no crises related to the loss of fresh water or soil degradation. Vegetal life presents a logic around producing habitability, supporting life, and this seems to be present in the willow basket weaving communities in Denmark. Their growing of willow and intimacy with the plant is a quiet activism.[1]

NOTE

1. Weavers are Ane Lyngsgaard, Eva Seidenfaden, Dorte Tilma, Dorthe Lynnerup, Gitte Rubek, Inge Lisa Rauhe, and Tim Palmer.

Chapter 5

Ash Tree Worlding

Elizabeth Oriel and Anna Perdibon

Thou owned a language by which hearts are stirred.

(John Clare, *The Fallen Elm*)

This volume engages with how vegetal life influences humans and other animals as place makers and as guides to contextual relations, such as in offering orientation to both time and space. In this chapter, ash trees and human relations with them are central actors. Ash trees, with 65 species in the *Fraxinus* genus, have been sources of rich stories, myths, medicine, and lifeways in many cultures that speak to the wide berth this tree engages with. The World Tree, a primary world maker and holder in Norse and Celtic mythologies and sense-making, is an ash tree; yet in the last decades, ash trees are dying across the United States and Europe. We explore how relationships are hewn with them, engaging their personhood and their influences that situate and orient humans in subjectivities and ontologies. This situating and orienting forms ethics.

Through extant literature and interview data, a kind of amber emerges, or resinous material from tree sap, that is made from diverse knowings across myth and folklore, ecology, and embodied relationships as ash trees are dying. Ash-human stories and relations are certainly multi-directional, though ash qualities and influences precede humans (with European ash dating back 11.6–5.3 million years ago) (Hinsinger et al. 2013). At the same time that plants are (re-)gaining recognition as persons through Environmental Humanities and Religious Studies scholarship (Hall 2019; Lawrence 2022), many plant species are dying in the current extinction crisis. This juxtaposition of recognition of plant persons with death intimates a response of kinship and a responsibility to others.

Threats to ash trees come from a feral insect in the United States and a feral fungus in Europe.[1] This chapter is a pausing, a recognizing of who these vegetal beings are, a taking note of ash tree deaths, a chance to dialogue and listen to one's responses. Emotions are allowed, are called for, are inevitable. Tree deaths help call attention to this sixth mass extinction event, to the dying of the world we know, but also to the personhood and stories of the arboreal being that is languishing, holding a whole world. As Van Dooren states about the loss of crows in Hawaii, local people say they have "lost the most intelligent and charismatic component of the forest" (Van Dooren 2017:191): though not typically perceived as charismatic, ash are known to be intelligent in the mythic sense as worldmakers and as protectors, especially in enmeshed relations with humans throughout history (Parker 2021).

A tree is in dialogue, in interdependence with many beings and forces. A tree is not just an organism but hosts a meshwork of lifeways, providing structure and sustenance for lichens, mosses, fungi, birds, and much more, and hosts meshworks of stories, experiences, and uses. As the Vegetal Turn reveals, relationships with plant beings can be complex affairs, as plants have ecological, medicinal, sacred, economic, gastronomic, and social roles and relations with humans and other beings. What modern industrial culture has denied and attempted to erase in many places is how plants and other beings interact with the non-visible world, with a complex cosmo-ecological worldly structure that has a geography in which trees often are access points. For example, as the World Tree, ash trees are holding many worlds together, mirroring their complex benefits to human lives.

This chapter imagines ash tree social influences through subjective, experiential, ecological, and mythic terrain. Social relations with ash are built up over time, as they are with humans, and require mutual personhood and relational ethics. From this position of shared terrain, we ask: Who are ash trees as persons in certain human experiences, which is an ethical question as they are losing out in the Anthropocene's fraught times? And how do ash trees influence humans in their orientation to place, time, and self? This position of holding the worlds, of surviving the end of time, and of connection to the origins of human life, positions ash as axial to both space and time, at least mythologically. And in the interviews presented in this chapter, ash do influence humans along these thematic lines. Their deaths call these human-tree connections into special illumination. As with other chapters in this volume, affective ecologies and intersubjectivity are modes of sociality across plants and humans.

Extinction studies offers perspectives on ash deaths by posing the themes of extinction as collective death as well as individual death, the importance and loss of intergenerational inheritance, and disruptions caused to the mutual generativity of the world (Bird Rose et al. 2017). We explore how myth, botany, folklore, medicine, ecological roles, intimate human-ash relations, including storytelling, art, and performance, speak to ash lives and occasionally how these ways of knowing mirror one another. This form of knowing and engaging is closer to how humans know other humans, and thus aids in perceiving trees as persons, as relatives, friends, and teachers in a phyto-human social milieu. We ask interviewees about their sense of ash trees and cosmology, and what it means to them that this tree that holds many worlds is losing ground. Interviewees are a diverse group of ecologists, storytellers, environmental advocates, farmers, and ash enthusiasts, offering guidance in respectful relations toward plants' living and dying.

In these shifting times of ecological collapse, we have much to learn from plant lives and stories. Ecological collapse can be viewed as resulting from disengagement and abstracted relations with the vegetal realms. Perceiving plant personhood is a response to the objectification and instrumental approaches that undergird the industrial food, medicine, clothing, and housing systems reliant on monocultures, plantations, and deforestation.

> Living in a world underpinned by the plant kingdom, the existence of plant persons is incredibly important for discussions of interspecies ethics. This acknowledgement of plants as persons is based on and in turn strengthens the recognition of plants as kin. Indeed, personhood is expressed and galvanized within specific kinship relationships between individual plants and humans. (Hall 2011:100)

From the perspective of an Indigenous tribal member in Brazil (Person commun. 2019), plants are more than persons; they are masters.

Plant personhood suggests a different approach to the social sphere, broadening past the humans to a multispecies and more-than-human domain. This is a situated domain, emplaced in soils, microbial relations, and interactions with sunlight. Maps of this social world are storied and possess many worldviews coexisting. This geography of relations to ash tree lives is a liminal space, where hybridities hold, and where the body of the world, body of trees, body of the land, and body of the mind and of the collective are shared terrains. More than bodies not having distinct boundaries, this land is also infused with mind across all beings and place. As Guattari in his *The Three Ecologies* noted, the ecological, subjective, and social are mirrored; Gregory Bateson, in his *Steps to an Ecology of Mind,* argues that the mind is not contained within the skin or the brain, but that mind moves through the world and through persons, such that mind and body are not distinct. Echoing this is Dorian Sagan, who writes that Earth and all beings "exhibit natural purpose—even a kind of mind" (2023:6).

Given this distributed mind, how can humans access or engage with plant personhood? Is communication possible? And how is plant personhood different from human personhood? Plants communicate across individuals with volatile organic compounds as their alphabet. Listening and responding to different voices, nonhuman voices, and even plants that have different perspectives is a form of multispecies conversation or dialogue, drawing on Russian philosopher Bakhtin's dialogism or world as dialogue, which both reflects the nature of reality and is a way of living across differences. These dialogues that emerge in plant stories are one route Matthew Hall offers to engage with plant sociality and personhood (2011:161). Living amidst kin suggests engagements such as recognition, communication, and learning from and about others, which also requires responsibility as kinship is guided by ethics. As Marder's philosophical work (2013) asserts, plants are different in how they hold no delusional distinction between self and other, as humans do. Marder says that plants are teachers to us in the sense that they don't restrict themselves into stable identities, as they respect alterity.

This chapter learns from ash in a kind of vegetal biography and in a eulogy to a tree teacher. Humans can come to know ash as one would another person, from perceptions of ash trees' personality traits, preferences, gifts, histories, and challenges, all filtered and enmeshed through human subjectivities. Ash in Britain has been a healing tree, and in West Sussex, it was known as magical (Mabey 1996:326). One interviewee from our research described how the World Tree is not a tree of knowledge but a tree of *life force*, ash being the sinews or connective tissue of the world. This essence of ash as life force echoes through ash stories. Ash tree sap is called "manna," like manna from the Bible, which is a heavenly substance and sustenance, similar to life force. Ethnobotanist Enrique Salmon writes that tools made from ash wood feel *alive* (Salmon 2021:26). Again, this theme of life and life force is a constant trait of the ash, something that is found in traditional cultures and mythologies across the two shores of the Atlantic. As mentioned in this volume's Introduction, *what we argue here is that by learning and experiencing qualities and aliveness of a tree, we can acknowledge its personality, its being as a person.*

Here we present a synthesis of seven semi-structured interviews conducted with those who have relations with ash trees, interlaced with folklore, stories about ash, myth, and ash ecology. Interviewees were chosen for having experiences with ash trees. We gather here and create bridges between Indigenous North American, traditional European, and Western scientific knowledge, stories, practices, and perspectives. However, we had the privilege to dialogue with two Anishinaabe interviewees, while the others are British, American, Scottish, Irish, and Australian. We analyzed interview data by first finding themes and then coding for themes, which led to interpretation and findings. With our Indigenous interviewees, we reciprocated for their time with an offering. This chapter begins with mythologies and perspectives about ash as the World Tree; then describes their relations with hydrology and related ecological contexts; next, we discuss practices and uses of ash wood which reveal more about personality, then death and renewal, and finally embodied and personal relations, including those revealed through art and storytelling.

MYTHIC PHYTO-HUMAN RELATIONS

With ash trees, their mythic roles are forms of influence in situating humans in an ancestral reaction with trees, circumscribing one's contextual origins and responsibilities. Ash trees are viewed as the source of life in many cultures. Mythologies about trees can be perceived as related to, or originating from, the distant past, yet these myths speak in the present day to interviewees as felt, embodied relations that shape orientations to space and time. Many world mythologies portray a World Tree or Tree of Life, which is conceived as the vital source of all existence. With its branches, the tree stretches into the sky, and with its roots, it reaches the depths of the earth. The tree is not only the cosmic pillar of the world but it is the main mediator between worlds. With its cycle of growth-death-rebirth, the tree embodies the energy of life and its ever-renewing power. In this way, ash trees are the axes of place and time; they begin the stories of life and situate them.

In Norse cosmology, the World Tree is Yggdrasil, which was likely an ash, though in some accounts it is a yew, and holds a cosmic geography of time and space. The *Edda* epic tells us that the ash is the greatest of all trees, the *axis mundi*, around which the gods gather in assembly and establish judgments every day. Its three roots sink deep into subterranean dimensions, with one of them reaching into the well of wisdom, a holy well whose waters are sacred and the source of all knowledge. On the top of the tree nests an eagle, a squirrel runs up and down the trunk, and deer eat its leaves. The base of the tree is inhabited by snakes and by the Norns, who determine human destiny at the moment of birth (Hall 2019:48–50). The ash is "the grandmother of all trees. It is all trees in one tree," says one of our interviewees, Gordon MacLellan. And in Norse mythology, it is the tree that "bridges the worlds"; as such, ash is the sinews, the connective tissue between worlds.

The word "Yggdrasil" can be interpreted as "the horse of Odin," a name that refers to the story when Odin hung from the ash as a voluntary self-sacrifice to obtain knowledge and wisdom in the form of the runic alphabet. (Memory Paterson 1996:149): "I know that I hung from a windy tree, nine long nights, wounded with a spear, dedicated to Odin, myself to myself, on that tree of which no man knows" (Edda Havamal, stanza 138). Such a practice of surrendering oneself to a different type of knowledge is not only a traditional shamanic practice of ritualistic death but can also be read as a way to learn from trees and from their rooting into a specific ecology.

According to Harriet Sams, the ash tree "invites us to explore what it feels like to have your skin tensed, and to wipe yourself with ash bark" as a way to "re-enact a somatic awareness," a form of relating with trees that reconnects to the life force that holds the world. In that embodied and spiritual experience we discover, Sams asserts, that the "tree of life is constantly there." Sams' experiences with ash can be viewed as intersubjective, as in a shared rhythm or manner of living (from Merleau-Ponty) across species and also as affective, such that what ash trees project is felt in her body. Another affective and intersubjective response is from MacLellan, who speaks of trees' perspective on life: "trees do not panic, they simply are and breath. They just live, until they die. And time, with its cycles of life and death, is a mysterious affair." This empathic practice is key in phyto-human social worlds and to "becoming with" as Donna Haraway speaks of what makes us human (Mentore 2012; Haraway 2008).

Long-held stories speak in the language of intersubjective states that open plants' lives to human awareness. Ash trees appear to be trees of life in an intimate and very practical sense in one Anishinaabe legend called *The Black Ash Basket*. In this story, the elder Black Elk, facing the end of his life and concerned for the well-being and future of his community, called for the Creator's help to heal his people's restlessness and to leave teachings about providing for their families. Black Elk was gifted with the following vision:

> When Black Elk died, the people were supposed to burn his body and then bury the ashes that remained in a special place. Out of the ashes would grow a special tree. The people were to watch over this tree and protect it from harm. This was to be considered a sacred thing. When the tree matured, it was to be cut down, and the growth rings were to be removed. (Benedict et al. 2014:142–143)

This story speaks of reciprocity inherent in the wheel of transformation: from the ashes of an elder, a sacred tree is born, a tree that will be turned into objects that will sustain the lives of humans. Indeed, the black ash is a tree of life in the embodied sense that human life depended on and could be supported by it. The basket-making process—from removing the rings to preparing the strips until the actual basket weaving—was shown to the elder in his vision, as much as the appropriate behavior in taking the life of each black ash. As Robin Wall Kimmerer stresses, "traditional harvesters recognize the individuality of each tree as a person, a nonhuman forest person. Trees are not taken, but requested. Respectfully, the cutter explains his purpose and the tree is asked permission for harvest" (2015:114). Accordingly, it is a knowledge that has to do with learning how to be human, to live life following the principles of patience, care, respect, and reciprocity, and to relate to trees as persons. Such knowledge of Indigenous basket-making is currently being renewed in the United States: an ancient traditional practice that is coming back to life with the very act of weaving reminding us of the interwoven quality of all life that is apparent in their ecologies explored next.

WATERY AND ECO RELATIONS

For some ash tree species, such as black ash (*Fraxinus nigra*), their geographies are watery and liminal. As Robin Wall Kimmerer (2003:15) writes, trees are tall and rigid because of their vascular systems that conduct water, speaking to plants' structural and functional relations with water. Black ash are riparian trees in parts of the midwestern United States and Canada, growing well near rivers, lakes, and in flood-prone areas. They manage their own hydrology to withstand floods and droughts as well as the hydrology of riparian woodlands and lakes. Ash leaves are important food for frogs and tadpoles, who live in lakes and puddles. Invertebrates falling from tree branches provide most of certain fish species' food, and leaves falling in rivers form the basis of the riverine food chain. Riparian trees stabilize river beds and lake edges against erosion, and protect lake hydrology in the midwestern United States, according to the Fond Du Lac Band of the Lake Superior Chippewa who grow wild rice in lakes and are land stewards (Nature Conservancy 2021).

Though European ash prefers moist soils, it has the ability to tolerate drought better than some other trees. It is known in England to maintain green leaves during drought when other leaves turn brown, providing food for hungry livestock. This ability of ash trees allows humans to withstand cycles of plenty and loss. Two compounds, mannitol and malate, give ash trees water-stress tolerance (Guicherd et al. 1997). Manna is a term for sap from various tree species, and the Biblical manna that God provided to the Israelites may have been sap from the Tamarisk tree. An interesting correlation is that ash sap and the compound mannitol are being studied to treat Alzheimer's and Parkinson's disease, as well as cystic fibrosis and intracranial pressure, acting as an osmotic diuretic (Better et al. 1997). This is another watery and fluid dynamic.

Ecologically, multiple ash species in North America and the European ash (*Fraxinus excelsior*) provide important habitat and food for various other creatures. The alkaline pH of ash bark provides excellent habitat for moss and hundreds of lichen species. Ash forms arbuscular mycorrhizal symbiotic relations with fungi that enter tree roots in the ground, providing nutrients between tree and fungus (Brundrett and Tederson 2020). Black ash alone supports wood ducks, wild turkey, cardinals, pine grosbeaks, cedar waxwings, and yellow-bellied sapsuckers, with habitat and food (such as the sap being of interest to the sapsucker) among others. Many mammalian species, from meadow voles eating the seeds, white-tailed deer eating the foliage, to silver-haired bats nesting, benefit from ash trees. Ash are key species for land snail diversity and for threatened ferns and lichens (Parker 2021). The European ash loses its leaves early, which supports the growth of dog violets, wild garlic, and dog's mercury.

Ash trees grow in diverse conditions, but ash dieback and the emerald ash borer are having profound impacts on biodiversity and on relatively rare floodplain ecosystems. As a keystone species in floodplains, their demise threatens the survival of vulnerable associated species. Both ash dieback (from fungus in Europe and the United Kingdom) and the emerald ash borers (beetle in United States) kill trees by strangling vascular systems, which mimics strangled hydrologies on the land. Floodplains and riparian zones are threatened ecosystems, having been terraformed with dams that control river flows and drained for development, which ultimately alters riparian species composition.

SPACES OF IN BETWEEN

One theme that emerges from ash tree ecologies and experiences in terms of orienting to both spatial and temporal realities is in relation to thresholds and transitions. Ash tree species in the United States such as Black, White, and Carolina ash inhabit riparian zones that are transitional threshold spaces between water and land. Ancient Celtic people thought of the areas where two realms meet as especially magical; ecologically, these liminal zones are unique in hosting high rates of nutrients and biodiversity (Turner et al. 2003). In the Norse myth, Yggdrasil holds the nine worlds: its spatial structure forms the transitions between the worlds. The World Tree embodies this spatial threshold, and yet ash is also present in temporal thresholds in rites of passage in human existence. Ash was regarded as a protector of newborn babies and mothers. Its boiled sap was given by midwives to newborns, a drop in their mouths as the first substance they tasted, gifting the baby its powerful vital force (MacCoitir 2003:122). Accessing healing from ash trees took place by passing a baby or child through a split ash trunk, which was then wrapped back together. Ash branches have been set on cradles to protect newborn babies. The ash tree energy helps with transitioning, birthing, and dying (Harriet Sams). In terms of death and temporal endings, in some accounts in Norse myth, the World Tree may survive the end of the world, providing a threshold and shelter for a single man and woman to repopulate the world.

PHYTO-HUMAN CULTURAL RELATIONS

Ash trees have had enormous influence as they have been the workhorses of rural life, with wood, leaves, and sap all providing substance and sustenance for humans and livestock for over 5,000 years in Europe with timber, leaves, sap, and bark (Parker 2021:114). The uses of ash date back thousands of years, and knowledge of the trees, how to interact with them, to learn their signals, and the uses of their wood and sap comprise European intergenerational heritage. The wood has been central to economies with spears, bats, coaches, poles, and lobster pots and retains minimal moisture.

Ash resides in the Olive or Oleaceae family along with lilac, jasmine, and privet. The word "ash" derives from the Old English *aesc,* meaning spear, and not surprisingly, the wood makes excellent spears due to a toughness that absorbs impact well, being paradoxically both tough and elastic (Parker 2021:115). Black ash wood has been used for excellent lacrosse sticks. Ash trees grow two different types of wood during the year, one more porous in early spring and summer, and then more dense and tough in late summer and fall. This quality of duality seems present in many aspects of ash lives, as a tree related to life and death, to everyday uses yet with strong mythic roles.

Though ash is still an economic and high-utility species, the loss of traditional practices and knowledge is vast, and the loss of millions of ash trees is a loss of intergenerational heritage where that knowledge remains. Ash tree qualities can often be often best recognized through practices. For example, ash trees are known in Sicily to love music by those who tap the trees for sap or manna, and if the cicadas don't sing, they say the trees won't produce great manna (Rupra 2020).

What was once perceived as a tree of healing and magic is now seen as a fast-growing but second class, weedy timber tree that colonizes open ground quickly (Mabey 1996). Perhaps this change of perception, this loss of recognition of their unique qualities and interrelationships with lichens, cicadas, and others, is at the root of ash tree deaths. With the hollowed-out relations within the local, as Latour (2018) writes, with a lack of attention, familiarity, practices, and engagement that regenerate the world (Ingold 2020), the life force of ash breaks down. As Madeleine Collie, who led walks in the Kent countryside said, many university-age students have trouble viewing a tree as a being that can have a disease. This presents a lack of ability to participate in intersubjectivity and interspecies socialities. Collie would walk through the dappled light created by ash's compound pinnate leaves and tell them this quality of light will be gone with ash's demise, as the large-leaved sycamore takes over. Her point speaks to one of the ash tree influences and affective roles with humans through their quality of light, a quality that is poorly appreciated in a world where economic gain is the essential value.

GEOGRAPHIES OF DEATH AND RENEWAL

Loss is such a natural part of life and yet at this moment, the world faces loss of life and species at such a rapid pace. Ash trees are not only dying at this time but also have much to say about death and renewal. The two ash tree extinctions are precipitated by a fungus (Europe and United Kingdom) and an insect (United States), yet are caused by capitalism's feral processes, undoing the relational situated co-evolved patterns that maintain system integrity. Ash dieback is a fungal infection caused by a strain (*Hymenoscyphus fraxineus*) from Asia. The spores travel through the wind, even traveling across the English Channel, and land on ash, entering the tree through leaves and stems. They enter the phloem, xylem, and pith, and thus spread throughout the tree. Ash dieback has been deemed untreatable and is likely to kill 90–98% of ash trees in the United Kingdom (according to government figures), and a similar number in parts of Europe. Similarly dire and in a similar timeframe, the emerald ash borer, a wood-boring beetle that also arrived from Asia, was first detected in 2002 near Detroit, Michigan, and is decimating ash trees in the United States. The beetle can fly up to 5 km a week, spreading devastation quickly (Taylor et al. 2012), and has no natural predators, no co-evolution specific to place, and could lead to the extinction of 16 ash species, 8 billion trees in the United States. The ash borer came over on wooden shipping pallets which transmit numerous diseases and pests along with the cargo. These globalized solid wood pallets are remaking forests, and the failures to treat and prevent this spreading from wood pallets land on local communities and homeowners who bear the brunt of tree removal costs while local woodlands change in their makeup (Weiss 2021).

Ash trees, which offered access to language in the runes for Odin, have a hard-won reputation as trees that will get you through hardship, drought, and disease. The Norse myths tell that Yggdrasil will survive the end of time, Ragnarok. Losing this enduring one, the rescuer, is a remarkable loss of eco-cultural inheritance and genealogy. Darwin wrote that species and language are both genealogical at their core (Grosz 2004). The ash as World Tree was an inheritance from an earlier time, as the supposed author of the Prose Eddas, Snorri Sturluson (1179–1241) lived in a landscape with few trees since most forests in Iceland were cut within decades of the island being inhabited. These inheritances for the dark ash bud moth, which depends completely on ash, and for the Yggdrasil stories are perishing. Living without ash will surely impact human relations to World Tree stories. The loss of ash mirrors our world ending as we know it, as climate scientists describe what lies ahead in the next decades. Tahnahga Yako, a Mohawk and Taino interviewee who is also an Anishinaabe tradition holder, says, "the ash trees are reflecting to us the sickness in the world. They are infected and we are affected."

However, trees undo our categories of life and death. In an old-growth forest, living and standing trees have only 5% living cells, while 20% of a dead tree is living tissue (Luoma 1999). This approach to life as residing well in the shared plentitude in death and decay leads one to an alternate view of self and other, a kind of sacrifice of self for others. And yet, with extinctions which are a collective death, grief is experienced by each differently; "there is no singular phenomenon of extinction; rather extinction is experienced, resisted, measured, enunciated, performed, and narrated in a variety of ways to which we must attend" (Bird Rose et al. 2017:2–3). "Extinction . . . .is never a generic event and is always a multi-contextual phenomenon requiring multi-disciplinary modes of encounter and understanding" (Wolfe 2017:viii). Collie describes her witnessing ash trees dying around her in Kent, United Kingdom:

> ash trees die slowly and they tend to go quite white, when they are dying, they sort of somehow bleach a little bit; they look a bit ghostly because they lose the leaves around the top of their crown. And the bark is whitish gray.

Embodying extinction resides in witnessing changes in color that emit a different gestalt, as she says, a ghostliness.

A general lack of connection and knowledge of the more-than-human world is mentioned often in the interviews in connection with ash tree demise. In a state of disconnection from the world, industrial humans can live in fictional worlds, where detachment is the norm. In this way, we live a parallel dream that prevents us from experiencing the interwovenness of life. Such a lack of embodied knowledge of trees' lives generates a lack of care. When a tree or species suffers illness and slowly fades away, this passing can go almost unnoticed, mirroring the lack of an appropriate response to the current extinction crisis. According to Sams, "it's like watching a family member with a debilitating disease and they're just fading away and nothing will make a difference." Indeed, even if we know that we are going to lose the ash, we still keep moving on with our usual lives, perhaps unconsciously forgetting about the trauma of losing a major part of what held our world for centuries. Of course, "it is painful to be reminded of what we love and are losing, or even to be awoken to the fact that once we did love our trees, our rivers, our landscapes" (Sams). This is why recovering and restor(y)ing relationships with ash trees and their stories is vital.

LANDSCAPES OF EMBODIED RELATIONSHIPS: EXPERIENCES AND ANECDOTES

Ash tree influence is so close to the bone that many stories situate human bodies as originating from trees. Some old European stories tell that the first humans were carved from the ash tree, tracing direct human-tree kinship. Greek myths describe the oak as the maternal ancestor of humanity and some mention how Zeus formed the human race from ash trees. The ancient Greek poet Hesiod tells us that ash trees are our ancestors. In his *Works and Days*, he wrote that humans sprung from ash trees, offspring of the ash-nymphs, the Meliae (Hall 2019:71). The introduction to the Edda reports the Norse gods Odin, Honid, and Lodur created the first man from *ask*, the ash, and the first woman from *embla*, the elm tree. In the Irish and Celtic contexts, belief in the arboreal origin of humans is apparent in the names of several ancient tribes or individuals who named themselves after native trees (Zucchelli 2016:176–180). Widely inhabiting the northern hemisphere, similar stories about the origin of humans from the ash appear in Indigenous North American myths as well. According to the Wabanaki, a confederacy that joins five Algonquin-speaking tribes in the Northeastern United States and southeastern Canada, humans were created from the Black ash tree (*Fraxinus nigra*). As Collie reminds us forcefully, "if these stories tell us anything, it is that we once revered trees without separation, learned to think through language alongside them, and perhaps in a distant ancestral memory, felt their flesh as our flesh" (Collie 2021:162).

The ash tree is an intimate relational partner that shapes how one experiences the world. One example is in the words of Madeleine Collie, who directed The Ash Project in the United Kingdom., describing ash trees as,

> having a special presence. They are tall, elegant, and gentle somehow. This has been recognized about them for years. They are the Venus of the Woods and ash trees are known as a cure for loneliness. I was interested to read this about ash trees years after I worked with them, and I thought, "that is how I experienced getting to know ash trees." But I am also wary of assuming to speak for the ash. I know that our mind and our consciousness is shaped by plants in a way I didn't know before. When you are able to recognize another species, it profoundly changes you. Ash is a very special plant which is why it is the World Tree.

Again, intersubjectivity plays a role in the depth of her perceptions and care for ash. Collie's experiences working in an organization devoted to protecting and celebrating ash trees offered special access to refine her perceptions and knowledge of ash's influences.

Ways to relate with ash trees as other-than-human persons emerged in our conversations. While building relationships with individual trees regardless of the species through weekly or daily walks in the woods of Derbyshire, MacLellan often sits close to an ash tree in a silent meditation. In his words, being with an ash tree, "is embracing the opportunity to share a moment of time with a friend. I am sitting next to it, and I do not ask questions, partly because it always feels rude." In his view, ash trees often embody a real stillness and gracefulness, a particular energy that belongs to the ash. Yako explains how in her tradition, there are actually seven directions, not just four, and the seventh is "the fire within, where we learn how to move from love and compassion. Ash trees and all trees have this seventh direction also. People need to be in touch with their own seventh direction and acknowledge the interconnectedness before they can be with ash people."

Dancing, painting, weaving stories together with, for, and about trees are all ways to bring us back into a reciprocal relationship with the living land, with the broader-than-human community we call nature. In Val Plumwood's words, the arts are able to *re-animate* the world, as they allow nature to speak in the active voice (2013:441–453). Stories, performances, and visual art are all powerful ways to create new stories, drawing from old ones, and to get to know trees' lives and stories:

> I think storytelling is utterly crucial. And the reason why I think that and I encourage it in so many ways, is that we are collectively somnambulant, we just have forgotten that we were in love with the land and with its beings. That is because we have lost our stories. And we have lost our direct connection to the land. [. . .] People just do not talk to trees anymore, do not hear the sounds and the language of the land, and therefore they do not care. (Harriet Sams)

Stories not only hold a wealth of traditional knowledge and teachings to carry a respectful life, but are also the "fabric of our humanity" (Salmon 2021:11), they teach us who we are and how to root in a landscape. According to Robin Wall Kimmerer, "language is our gift and our responsibility," where writing becomes "an act of reciprocity with the living land" (2015:347). This deep understanding of story and art at large is shared by Yako, who asserts the generative power and vital necessity of story and artwork as ways to formulate a relationship, to explore an experience with another being, and to inspire a deep awakening. In this light, old stories, myths, and contemporary stories of personal, intimate relationships with the ash trees and their dying, as much as dancing, painting, and other creative expressions, are all precious gateways for bringing us back into a relationship with the living Earth, for reconnecting with the cycle of transformation that is life, and for learning how to take care.

Through drawing, writing, photographing, and walking, the people involved in the *Ash Project*, based in the United Kingdom, generated a somatic connection to the trees and the land, with the curatorial practices contributing to the creation of a soil where care can manifest. For MacLellan, while dancing in the forest with a mix of human, branch, root, and trunk, the borders between species dissolve, and trees become dance partners. As a storyteller and educator, MacLellan feels compelled to be the bridge between the nonhuman people and humans. In summer 2020, during the lockdown, he invited people into the woods of Buxton to share impressions of the forest and of ash dieback. A sense of ominous change permeated the time with trees. The experience brought to life a dancing puppet and a poem. The poem is a conversation between the ash trees and the humans visiting the woodland, their stories and their dying:

We are the trees
Who grow the leaves,
Who hold the birds,
Who paint black buds on grey fingers,
Who rattle our keys,
Who lift our trunks,
To raise the clouds.

We are the trees
Who shed slim grey dresses
For coats of green and brown and lichen,
Who comb the air,
Who dance in the wind,
Who grow hope from seedlings.

We are the trees
Who dangle icicles,
Who drip rain,
Who spin mist into dreams,
Who grow beards of moss,
Who raise the sap,
Who foster the greenfly
To feed the nestlings.

We are the trees
Where the owls nest,
Where the bats fly,
Where the spiders weave,
The woodpeckers drum,
And caterpillars feast.

We are the trees,
Where mushrooms grow,
Where buzzards scream,
Where badgers sett,
Where worms wriggle,
Where rabbits burrow,
Where foxes earth,
Where squirrels drey.

We are the trees
Fraxinus, that's us,
Fraxinus excelsior
Fraxinus, the Romans named us
Excelsior, the lofty one.

We are the trees
Who dread the wind.

We are the trees,
Where the spores settle,
Where the fungus spreads,
Where the fingers wither,
Where the bark splits,
Where branches break.
We are the trees holding onto hope
In seeds and seedlings,
In long breaths held and
Hearts clenched against the dread.

We are the trees who
Grow the keys of hope.
(credit: "Gordon MacLellan for Stone and Water")

TOWARD A CONCLUSION

As one respondent shared with us, there is something profound and generative when pausing with a tree that is recognized in our ancestral stories as being a World Tree. It is not something merely symbolic, but rather a call to action: a call to recognize and reclaim these ancient stories as a place for renewal, where the scale of loss is a call to change and mend our relationships with the living world. A first step in this urgent transformation is to start looking at and engaging with ash trees and the other plant inhabitants as living beings with their own distinct personality, their own geography, and stories. Along with this is the recognition of ash tree influences on humans such that the trees and humans share a multi-directional sociality in intersubjective relational lines. Relational and artistic approaches can aid this pathway of shared origins, loss and grief, of care and renewal.

Ash tree geographies and influence form a confluence of mythic, embodied, and practical effects on human cultures, with numerous accounts of human origins in ash. Following traces or movements of ash in human ecocultures is a route of liveliness, vivacity, and connections, accessing sinews of the world. The mythic and practical forge together into a geography of shared access and persons across vegetal and other forms. In all these ways, ash trees offer human relationships to their context in their possible origins, in generous offerings during hardship, and in the nuts and bolts of daily life. The Greeks remind us that nymphs of the ash trees, the Meliae, were not separated from their plants but rather the "expressions of the sentience of their plants, the personified personhood to which humans can more easily relate to" (Hall 2019:152). Thus, they were interdependent, as we all are in the dance of transformation that is the world. What we suggest here is that the ash trees—trees that held a world, trees that are fading away—are embodying the sickness pervading our world, mirroring the end of it as we know it. However, in their self-sacrifice like Odin, ash trees will be waiting for a new time to sprout again.

NOTE

1. We use feral here as Anna Tsing does in her *Feral Atlas* to denote the process that beings go through in response to human infrastructure projects, as part of empire and industry (Tsing and Bazzul 2021).

Conclusion

Vegetal Influences and Meeting of the Minds

This volume offers perspectives about plant influences and agencies, and ways to notice and attend to them. As mentioned in the Introduction, my friend Terra Soma in Portland, Oregon, allows plants to influence place, by letting them settle and do their ecological work, repairing soils, and offering benefits to other beings, such as insects, fungi, plants, and animals. The dandelions are rooting around with their taproots, unearthing compacted soils, while mallow soothes dry soils. Plants seed into where they are needed and also where they are adapted to thrive. In this way, their thriving is a function of bettering or improving living conditions in their place. Though this process is upset by highly globalized and non-situated plant species that can opportunistically dominate others.

This quality of improving place is a possible entryway into the mind of the vegetal. A meeting of the minds across human and the vegetal is apparent in various forms in the case studies. Each part of a plant's world intersects with the whole system, in how plants foster habitability, improving possibilities and prospects for life to continue. I hear activists today speak in these terms, of improving possibilities for life. Plants, as generators of habitability and place, are guides to this kind of activism. They are beings of context, offering humans greater relations with contextual patterns and beings. One language of this contextual world is poiesis, a force of shared creation, a sensitivity that begets creation and lyrical language.

Across the five case studies, multi-directional socialities across humans, plants, and place are at play. These socialities are rooted for humans in attention to plant life, skill and practices in poiesis, and in the sensuousness of experiencing their qualities. Affect and intersubjectivity are ways to feel with vegetal others and to pay attention to impacts and effects. In each case study, plants' qualities are important to human culture and lifeways. Even when vegetal subjectivity is not necessarily recognized, their qualities open up a field of relations. And in the meantime, plants are responding to how they are treated, and if one takes care of their interests, they will repay generously (see chapter 5). The implication in this statement is of mutual personhood, mutual subjectivity. That they are aware and respond to how they are treated goes beyond how many in the West view their capabilities.

Vegetal life opens humans to their context in bodily relations, in experiences with one's land base, with larger landscapes, and with systems such as nation-states. These experiences take place across the so-called nature/culture barrier, helping to eradicate that delusional separation. The paradox is that those who live from and closest to their land base comprehend the interwoven qualities of nature/culture, yet they have so little voice in dominant society. There is often a potent and realistic perspective found in those close to the land, a horizontal and fungal viewpoint that recognizes how the Earth offers abundance in particularities of time and space. This contrasts with a scarcity mindset that drives modernity and material accumulation. The way water diffuses over a surface can be studied as a form of equitable resource distribution, for instance. Many examples of how water and forests construct worlds can be models for an abundant and equitable mode of life.

In part due to the sessile nature of plant life, their heightened abilities to sense their world and respond is a kind of perceptual expertise. This perceptual skill can be developed in humans through what American herbalist Stephen Harrod Buhner calls the feeling sense. Life works through feeling, he says, not thought. The feeling you have when you first enter a room, and you assess the atmosphere and may choose to sit in a location that feels best. A feeling sense has been proven statistically to work when observing nonhuman animals on farms. Francoise Wemelsfelder's (2001) research looked at whether animal welfare observers on farms can accurately assess animals through their body language, using their feeling sense. Her results affirmed this, finding with statistical significance that humans can accurately feel, through observing another's body, how they are faring. Yet, reading body language or reading an environment's feeling qualities are not especially valued as ways of knowing in Western culture.

Perceiving the world through feeling the physical world is at an all-time low, as screens have replaced lived experience for many. Videos of kittens replace actual kittens, and online porn replaces actual sex. Real in-person relationships fall away from a lack of attention. A dispossession of ourselves correlates to dispossession from forests, Robert P. Harrison argues in *Forests: The Shadow of Civilization* (1992). The word "forest" most likely comes from the Latin *foris* meaning "outside"; this etymology tracks with a sense of human life as existentially outside of and separate from forests or woodlands. This outsider status translates to loss on the ground, as forest cover declines increasingly each year. Worldwide, in 2001 forest loss was 13.4 million hectares, and in 2022, that figure almost doubled (World Resources Institute).

Yet despite these figures, house plants are having a moment of popularity. Americans spend about 90% of their time indoors, and yet relating with indoor plants has been associated with connection to wider ecologies and cycles (Phillips and Schulz 2021). In an article on houseplants in Atlantic Monthly, Garber writes,

> Potted plants have a quiet poetry to them, a whirl of wildness and constraint; they make the planet personal. I loved caring for ours. I loved noticing, over time, the way they stretched and flattened and curled and changed. I still do.

Research finds humans notice a calming effect from being in proximity to house plants that is similar to affects in outdoor green spaces. One aspect of plants' affects comes through their fractal patterns; viewing fractal patterns in ferns, for example, and the complexities in plant forms versus buildings have calming and restorative affects (Hagerholl et al. 2008). When asked to visualize a tranquil setting, many will imagine a field of wildflowers.

At certain moments in my own felt experience of plants, I notice a sense of contextual generativity in how certain trees and plants feel like doorways to another place or to certain stories. As a child, honey locust tree seed pods I found on the school playground in Chicago were storied artifacts that peopled imaginary games. In Massachusetts, red maples color a swamp in a red autumn glow against the blue sky, giving a sense of beauty so stark that doors open. In Sri Lanka, certain mighty fig trees have a bold yet humble presence and seem to be people. These experiences draw on a feeling sense that Buhner describes, and yet they open perceptions of trees as storied and mysterious.

The historical colonization of our ancestral ties to plants and their stories, knowledge, histories, and powers is a long one and much more pronounced here in the West. The climate and extinction crises, as well as economic and democratic crises, can all be viewed as emerging from an outsider position to the living world. Ecological degradation is inevitable as the modernist approach to the world eats the living world alive as it goes along. False narratives and perceptions of who humans are in the world are also part of modernity's grip, obscuring human responsibilities to the Earth. An alternative is politics and land use practices that are guided by planetary boundaries as signposts, as limits by sentient and earthly processes. Policies that support watershed integrity, for example, would mirror the centrality that water plays in earthly life and mirror watersheds' foundational roles in social organization. As eco-crises worsen, everything will depend on the health of one's land base, writer Derrick Jensen says.

A place to begin with this guidance by the living world and Gaian planetary boundaries is with vegetal life and how humans ally and partner with plants in co-generating places of mutual thriving. Jacaranda, willow, wapato and camas, ash trees, and plant actors in human-elephant relations and conflict have all shown themselves here as subjects with qualities and influences on animals. Throughout the specificities of each plant and their situated worlds, their influences become apparent in the complexities of their relationships and their abilities to bring humans into relation with their context. Temporal and spatial contexts become relational contexts in embodied time and in place. They bring the abstract into the embodied.

Plants orient humans to context, to time and space, and meanings related to place. Place is a meeting of time and space, as relations that form place are temporal and spatial. Place is a meshwork across time/space of memory, meaning, belonging, and co-generating of the world. Plants are also meeting places of time and space, and as Casey and Marder (2023) argue, are equivalent to place. They temporally and spatially show us our origins, as in Genesis with the two trees and with the ash as the World Tree; they reveal the changing of seasons that can be read on their leaves and on tree trunks; they reveal spatial orientations as in navigating the land and one's memory (more on this below); they constitute social worlds, as in wapato's world of Chinook-speaking human settlements and trade. Willow opens up weavers and growers to their context in the sense of loosening the grip of domestication that is a closing off of contextual relationships from view. Willow basket weaving also works through intersubjectivity across phyto-human bodies, as humans attend to rhythmic patterns of the plant and to the will of the rods. And in Australia, the fraught acts of settler colonialism that remake the land to suit newcomers' needs, jacaranda trees remake place and organize humans in social events and in relations to beauty and belonging.

In general, plants live with great environmental flux, in seasons, in heatwaves, in epidemics of disease. These fluxes mirror the world today and plants' abilities to persist through adapting is a model for the Terrestrial, which Latour describes as a needed yet poorly defined systemic attractor. At this moment in time, the lure of what Bruno Latour (2018) called the "Out-of-this-World" systemic attractor (that is a way of being in the world that views no material limits imposed by nature and thus lives as if off the earth) is increasing. A pull toward the Earth, the real, rooted, Terrestrial, the vegetal, is a necessity for a shared future. This return to the vegetal for the West is a widening of meanings and recognition of relationships of interdependence. One challenge of the present is a divide in relations with the future between an even more mechanized model of reality and a nature-based model, and plants offer myriad complex approaches to thinking with living systems, with the Tree of Life.

Plants both offer situated diverse complexity and also universalism when they are removed from co-evolved situated relations. Certain species come to mirror extractivist practices of dominant capitalist culture. As in non-native *Lantana camara* in Sri Lanka or the feral fungus that kills ash trees in Europe, plants adapt and respond to systemic parameters. If the system is geared toward extraction as in plantations across the globe, plants can mimic this mode in universalizing place with one dominant species. Capitalist modes remove nature from culture, such that plant cultivation is not a poetic practice or poiesis but an economic and extractive one, unearthing and deracinating situated co-evolved relations that have kept each species in check. Without feedback loops across species and place that provide and maintain proportionality for each, some that are more opportunistic, like *Lantana camara*, colonize the land.

Yet how plants mirror human extractivism is a testament not only to the physical repercussions of monocrops, but also attests to what Gregory Bateson wrote as the ecology of mind, with evolution being not only a physical process but a mental one. Mind extends across all aspects of the world, thus not being bound to the confines of human brain structures. Ryan, Vierera, and Gagliano's volume, *The Mind of Plants: Narratives of Vegetal Intelligence* (2021) speaks to plants' consciousness and meetings with human minds. The subjective experiences described in the book bring plants back into the fold of meaning, in which plants' qualities are relational and impactful.

Across the chapters, there is recognition of changing plant presence and communities in many contexts and a grief for lost ash trees, loss of recognition for wapato's gifts to place-making, the native plants crowded out by jacaranda trees in Australia, and *Lantana camara* in Sri Lanka. The senescing of historical plant species such as ash trees, for instance, which are central actors, becomes natureculture losses, as these trees were connected to humans in numerous stories, medicines, and rituals, while playing crucial ecological roles. Plant extinctions and changing community structures bring novel ecological relationships to the Earth, new foundations. Attention to these losses, to changes on the ground, and to new systems emerging are aided by perceptual attunement to vegetal relations and time frames as they both build relations in place to thrive and undo relations.

Another possible meeting place where plant and human minds intersect is in spatial orientation, and how this correlates with cognition, as mentioned in the Introduction (chapter 1). I make a case here for how plant minds and human minds not only meet in various ways, but how plant minds structure the living world through structuring aspects of cognition, and in particular, memory. In the little-known book, *The Art of Memory* (1966), which has a passionate following, historian Francis Yates describes various mnemonic practices in ancient Greece. Living until 468 BC, the poet Simonides created a mnemonic system called the "loci method." An experience he had one evening after a dinner party led to this discovery. He left the party before the others; the roof collapsed and crushed all the remaining guests beyond recognition. He could identify each of the remains based on his memory of where each person sat at the table. From this experience, he developed a spatial theory of memory, which became a popular practice to memorize long speeches. The technique involves visualizing a symbol for each point in a text and situating the symbol at certain points inside an imaginary building or room. These spatial locations guide one's mind through the memory of a text. In the 1600s, Robert Fludd's memory theater is another spatial mnemonic practice.

These techniques highlight the inter-relatedness of memory and thinking with space and place, showing the interdependence of time and space. The ability to hold a thought over time relates to situatedness. Minds and cognition are spatially oriented, which correlates with Bateson's work on mind that extends over land and sea-scapes. Over the passage of time, space provides organization for thoughts and memories. Thus, memorizing a long speech is done by situating one's topics in a room, or even outside, in a forest. Trees in a forest are a spatial organization of the land, which are also cognitive structures, as Bateson argues.

What comes to the fore is a sense of the world as a nested set of organizations. Trees and plants organize the living world and thus set up spatial organization for human cognition and memory. Yet, plant lifeways and structures are also expressing other levels of organization in soil type and minerals, and these reflect geologic and hydrologic histories that form soil, mineral-rich sediments, and movements of water. Thus, plants reflect their ecological contexts in the ways they structure space and place, and this nested organizational flow extends to human minds and memories. Thus, vegetal lives influence human minds and lives in the contextual aspects of geologic histories, or geo-social histories, while also helping to structure and organize one's thoughts. Each part of the world connects to a vast web of organization. It is no wonder that Celtic people created an alphabet from tree associations, with the vegetal as foundational to language.

Though despite this intrinsic interconnection, industrial humans map their minds inside brain structures, not onto forests, and map their social organization into human categories, as in nation-states. Nations trump ecologies as social worlds, though plants are asked to play roles in nationalism. The maple leaf on the Canadian flag and the made-up story of George Washington cutting down a cherry tree to promote his reputation are examples. Yet, the nation-state as a top-down and rather arbitrary category that is not aligned with ecological realities gets in the way of prioritizing socio-ecological realities in which plants are agents.

Meanwhile, other species continue to provide all necessary elements for humans' lives. The roles of each being involve sensing, perceiving, responding, and adapting, and these have maintained and regulated the Earth's atmosphere and temperature for eons. *This self-regulation across the planetary system is a kind of mind,* Dorian Sagan (2023:xx) writes. James Lovelock and Lynn Margulis describe Gaia or Earth in this way, as a superorganism, with each contributing their own lifeways to a larger purpose, which is sustaining life and livability. The ecological roles (there are certainly other roles as well) that wetland plants play, such as wapato and willow, in removing toxins, stabilizing soils, providing food and habitat for their ecotone or in-between place, are complex and significant. This is why biodiversity is essential to maintain the system. And some species, like ash trees, foster hundreds of other species. They are ecologically considered keystone species, and many species, like wapato are not only "keystone" ecologically but also culturally.

SOCIAL LIVES, POIESIS, AND COEXISTENCE

Plants have their own social worlds, such as in sharing resources and communicating to warn conspecifics and others in a forest of insect presence. They also have social worlds in symbioses with other kingdoms, such as fungi in mycorrhizae and in lichens, which are symbioses between algae and fungi. And plants' social influences extend across animals and generate place-based relations in numerous ways. As seen in the jacaranda chapter (chapter 2), tree presence and their life cycles orient humans to place, nation-making, and identities connected to one's land base. Affect is a primary mode of plant influence, as affective lines move across species, contributing to a sense of belonging and the ways psyches are shaped by place. Practices in cultivation, gardening, and urban greening all are foundational to other relationships, as witnessed in human-elephant-plant relationships. Forests and floodplain vegetation both promote localized autonomies, as explained in chapter 1 in relation to Asian elephants and in chapter 3 in relation to wapato. These explications of phyto-human social worlds are attempts to view qualities and shapes of foundational relationships on which all human life depends.

Plants structure the land and structure space for animals, and in so doing, co-generate and engender coexistence. Willow is one such plant actor that improves soils by removing excess water and toxins, prevents fires, and provides food for numerous species. With willow's many cultural associations with humans, this plant is a world maker for others, especially in the climate crisis. Plants and human-plant relations mediate possibilities for coexistence and conflict for humans and elephants. For farmers and elephants, the extent, quality and location of forest cover are essential to maintaining autonomies for elephants, in which they have freedom of movement, access to palatable vegetation and minerals, and a sense of their territory. I spent time in central Sri Lanka, near Anuradhapura, where only small patches of forest remain. As such, locals attest to how elephants live like homeless individuals, moving from the safety of one forest patch to another. Expansion of agriculture along with dams and reservoirs has terraformed the land into an inhospitable place for nonhuman animals.

In the last 50 years in Sri Lanka, human-plant relations have lost the glue of poetry, or praxis with poiesis, involving sensitivity to materials and conditions. Poiesis, to reiterate, is a term that relates to making something new in the world, as in what a craftsman does, and also in the root of the word, to poetry. Farmers, only a number of decades ago, would sing poetic lines to their crops and to the elephants, as a form of protection and as part of maintaining interspecies and trans-species relationships. Cultivation is a form of phyto-human poiesis in bringing food into the world that overlapped with the use of poetry. Singing and poetry were relational and sensitive modes of relating to place, to plants, and to elephants.

In the "Poetics of Cohabitation," Ribo (2022) describes how oral poiesis in the form of song is central to cohabitation across species and has been central to many Indigenous practices on the land. Oral poiesis, he writes, is an eco-semiotic form bridging human and nonhuman worlds by sustaining worldviews, such as animism and totemism that maintain human-nature social worlds, through symbols in metaphor, analogy, and others. Play, attention, musicality, feeling, and remembrance may all be present, fostering systemic relationality. These features of song speak to the Indigenous leader who sang to the trees on Hampstead Heath, mentioned in the Introduction. His songs were not a pleasantry but a way of engendering the world. A quieting of these songs in the world equates with the loss of cohabitation. New kinds of phyto-human socialities are emerging quickly in surreal ways as the next section details, matching rapid ecological change, and these new structures may require different songs.

NEW PLANT SOCIALITIES IN NOVEL ECOSYSTEMS

Thinking of how plants structure cognition, it is possible with novel ecosystems arising everywhere, that vegetal minds are asking humans to take up a completely new form of thinking and memory. Walking on Santa Cruz Island (one of the Channel Islands) 18 miles off California's southern coast, is a case in point. Grasses and yellow fennel flowers on tall stalks clothe dry brown hills and contrast with the Pacific Ocean's medium-range blue. The aromatic fennel is not native, and in their abundance, makes the land seem and smell like a large spice tray. All the plants I can see from this vantage point are not native, but are Mediterranean, from the Levant and Asia. The Chumash people who lived here before Spanish conquest lived off abundant plant foods, such as the underground corm of blue dick, and bountiful fish. Since settler occupation and as a result of indiscriminate land use and overgrazing, this island experienced an ecological collapse.

I spot a rounded and lobed leaf that reminds me of a hand. This small edible fig tree (*Ficus carica*) has been taking over wetland and river habitats in coastal and central California. On this day, I am here to observe non-native plant species as I write descriptions of each for an art exhibit in Carpinteria. The environmental organization, The Nature Conservancy (TNC), which owns 76% of this island, is removing these plants in a somewhat controversial effort to restore native island oak woodland here. In fact, their goal is to create a pre-Columbian landscape with native plants and animals (Shelton 2004).

Fig trees may be one of the first cultivated plants; their fruit (or really a syconium, which are inverted flowers inside a pod) played significantly into the beginnings of human societies, yet now grows invasively on this Channel Island. Figs are just one "invasive" plant here, among chicory, bindweed, Monterey cypress, and more. Fig trees reproducing across California in their geographies of what we call "invasion" are part of the human-disturbed, colonial, invasive regime that led this island toward ecological collapse. As Maan Barua (2023) suggests, vegetal agency can be pernicious, as with the *Mikania* vine in India, which was introduced from South America as a cover crop for tea plantations. The logic of plantations becomes an exponential spreader of extractive logics across landscapes, as plants move and "invade." Fig trees are potent forces on the land, with their close relation to human cultures and the sacred, and yet now are spreading and crowding others out.

The distinctive 3- or 5-lobed leaves mirror human hands, highlighting how human and fig lives are intertwined. In Jordan, the common fig (*Ficus carica*) was cultivated 11,300 years ago and was one of the earliest known cultivated plants. Dried figs sustained travelers on long journeys. Gifting a fig tree sapling is common in Arab cultures. In Sri Lanka, where I studied human-elephant conflict, fig trees are considered the most sacred tree, sanctifying Buddhist temple courtyards and contributing to atmospheres with graceful limbs and leaves. Each temple fig tree is considered linked with the very tree under which the Buddha attained enlightenment (a *Ficus religiosa*), and yet these trees also play large roles in Hinduism, Islam, and Jainism. Fig trees are considered to be gods in India (Begum and Barua 2012), and the branches are often thought to be home to gods. Adam and Eve used fig leaves as clothes. In ancient Egypt, they stood at the boundary of heaven, where a goddess emerges to welcome dead souls as they enter.

Walking on Santa Cruz Island, where Chumash people dwelled until the last Indigenous inhabitants left in 1822, I wonder if I might see a native species that they called the "Magic Plant," from which the Chumash say they emerged. Though I search a bit blindly, plants have a role in myriad cultures' origin stories, though not for the white settlers in this land today. The United States has no common plant stories to connect them to place, except perhaps the fabricated one about George Washington cutting down a cherry tree. Though that is not a story that informs about plant qualities and sentience or kinship relations, as Indigenous stories often do. This island has been inhabited for over 9,000 years. Plant geographies have defined this place through Chumash livelihoods and stories involving plant species, and now these new, more invasive geographies are choking the native species to death. The fig, this tree of the gods, is now an "invader" in California, deterring native species and altering soil composition.

While plants move through landscapes, seeding and begetting offspring, they exert influences on a place, revealing deep earthly processes. Fig trees influence humans in India and Sri Lanka through an orientation to both time and place. For Buddhists, they connect people to ancient times when the Buddha sat underneath the tree, drawing an invisible temporal line of influence and invisible spatial lines in their central position in temple grounds. Culturally significant plants have these roles of orientation that are powerful and touch on geographies of identity and belonging. On Santa Cruz Island, figs contribute to a disorientation toward both space and time, as the socio-ecological native relations are erased and new cultural associations are not established. Co-evolution is a potent force across plants and humans, and these novel ecosystems are starting from scratch, with opportunism as a primary organizing principle, matching the capitalist extractive approach to the Earth. Yet, the legacies of ecocultures across humans and fig trees seem to be present still. Through human eyes, fig trees have been beings of plenty, forming a generative relation. I cannot help but wonder if these fig trees are here to offer food in what will likely be times ahead of food scarcity.

Despite these changing ecologies and controversies over creating a pre-Columbian landscape, poiesis continues to be central to multispecies world-making. My friend and colleague, Lisa Jevbratt, a textile and digital artist, created weavings from yarns dyed with invasive plant species on Santa Cruz Island, including fig trees (figures 6.1 and 6.2). These weavings form a poiesis for a new way that plants are working to structure space differently. She worked with around 40 plant species and showed these works at a new museum devoted to the Channel Islands in Carpenteria, California. These weavings are an offering to a future of plant-informed approaches to the future (figure 6.2).

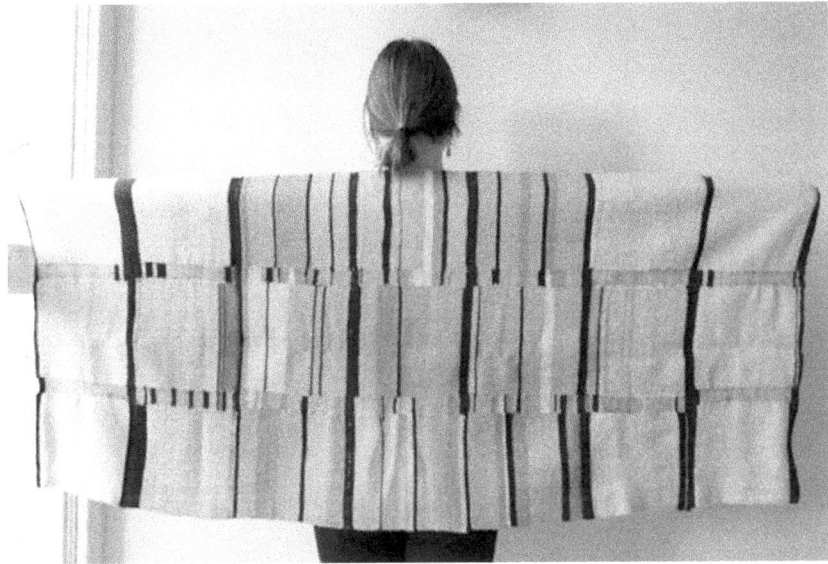

Figure 6.1 Weaving by Lisa Jevbratt as Part of the Project: Interlopings—Colors in The Warp and Weft of Ecological Entanglements. Yarn was dyed with Acacia melanoxylon, Eucalyptus, and Malva parviflora from Santa Cruz Island, California. http://rosebud.arts .ucsb.edu/~jevbratt/interlopings/. Lisa Jevbratt.

Plant life is perhaps the original aesthetic for humans, both in forming humans' approaches and perceptions of beauty and more generally in forms of perception. For Gregory Bateson, aesthetics is the way one perceives change in one's ecological context or habitat (Harries-Jones 2005). Trees and plants are our primary indicators of temporal change and also mirror the current climate crisis. The hillsides of charred and disease-riddled trees in the Pacific Northwest, just east of Sauvie Island in Oregon, speak to this moment of climate crisis. The trees tell it all when it comes to mirroring humans' socio-ecological position vis-a-vis the living world.

In their appearances, the vegetal speaks and influences humans. The vegetation in the tropics, with so much diversity, beauty, and majesty, had a great influence on Alexander von Humboldt, and he expressed sorrow for those in the north who will not experience tropical species. He calls on poets and artists to present their beauty to those who lack in-person experience.

> In the cold of the North, in the starkness of the heath, the lone individual can acquire for himself that which is being explored in the most distant latitudes, and thus create within himself a world that is the work of, and is as free and immortal as, his own spirit (2016:165).

Figure 6.2   Natural Dye Visualizations from the Project, Organized According to Their Hue. Appearances, aesthetics, feelings. Lisa Jevbratt.

Clearly, Humboldt speaks of internal relations with plants as equally vital as external relations. And these internal relations also have their influences.

Beauty is a powerful force that can be harnessed for various purposes. Resplendent jacaranda blossoms allow colonial settlers to feel at home amidst the beautified surroundings created by this tree, another non-native from South America. Somehow, native species created hardly a stir in newspapers, while the jacaranda blooming times elicit social events, marking the year in a flurry of purple blossoms and offering a sense of belonging and memory.

Vegetal aesthetics are affective, shaping experiences and emotions. The dappled light that ash trees create with their pinnately compound leaves (mentioned in chapter 5) is a placemaking aesthetic quality. Jacaranda blossoms excite and inspire people with poetic and visual art. Intersubjectivity, in which shared rhythms across beings form a bridge of experience, generates an interspecies shared experience of place, an empathic exchange. These aesthetic influences defy bold distinctions and categories. They live in the cracks of these divisions and open up possibilities for life to take shape.

Willow is an aesthetic partner and actor for basket weavers in Denmark. The willow case study calls up something unique in the volume—a physical, bodily intimate praxis and poiesis, a relationship felt in one's arm muscles that accesses the plant's will. The weavers feel willow's willingness and unwillingness in their muscles in pulling rods through and around other rods. Though many weavers use other materials for projects, willow is the central plant for most, and in the growing, drying, soaking, and weaving, the plant dictates its terms and negotiates how it will be manipulated. Willow's fast growth with new shoots every year that are long enough to be harvested, and its flexibility to be folded at sharp angles and looped around defies what seems possible from a woody plant. Willow's aesthetic is also palpable in its glorious, musty scent.

Bateson's concept of ecological aesthetics as related to changes in habitats applies to connections between appearance, beauty, change, and plant bodies revealing seasonal change while also somehow persisting across change. Trees reveal change and also defy change. Again, this correlates with the parts-to-the-whole relationship that can seem paradoxical. As described in chapter 4, Alexander von Humboldt left a legacy of holism in his approaches to plants, bridging natural and social science with art. Just as an ash tree is known in the realm of myth, medicine, ritual and magic, manufacturing, livestock farming, and hydrology, trees and plants defy imposed categories and instead speak to integration.

RECURSION AS A VEGETAL LEGACY

Knowledge and learning of and with the vegetal in this volume find heightened relevance in dialogues across ways of knowing. Ecologists, artists, and religious leaders can find parallels across their respective fields such that an oak tree's ecological qualities speak to Biblical references as well as to artistic ones. Critical Plant Studies provides a similar synergy of disciplines to meet plants' complex biointelligence. Plants are also knowledge holders in themselves and are also authorities on adaptability, habitability, and earthly processes. My friend Terra trusts their authority in her garden.

How vegetal beings grow is a learning process, an epistemology, as described earlier, a kind of gestural learning from how shoots grow and spiral, reaching sunward. The recursive process moves forward by moving back, going forward into the world and then renewing one's history with this new information. The reflexive mode of learning seems to be what plants offer, in which feedback loops provide information and reality checks. Making a willow basket is a recursive process as the feedback loops consistently reveal how willow responds to touch and warmth, and one can adjust with new knowledge of what does and doesn't work. This process could become a model for how to interact with one's land base, with trial and knowledge gained, adjustments, in an iterative process.

For example, when too much forest is cut down in Sri Lanka's southern dry zone, a negative feedback loop leads to elephants' greater reliance on crops, which, in turn, degrades farmers' living standards. And with a loss of forests that had fostered autonomies, feedback loops lead to less autonomy, more technological fixes needed, more IMF funding, and more debt, and a trans-species culture based on dependency. Feedback loops, if regarded well, are processes of natureculture politics and draw humans back to the realities on the ground of the living world they are seeking to control and to overcome. Thus, recursive learning modes instruct in staying true to diversity in place and history, while shaping multispecies forms of governance.

MULTISPECIES POLITICS

Here I return to the beginning scene, to the fragmented spaces in southern Sri Lanka, where vegetal agencies are removed from situated place-based associations. Elephants here have little to eat in a degraded national park that has been overgrazed, while sugarcane grows in a plantation that will create more potent alcohol to numb the difficulties of this industrial life way. Farmers rose up in 2020 and called for different elephant policies that give them more ranging areas, which would alleviate the conflict. Their demands speak to a multispecies politics, in which elephants are actors on the land and have needs, interests, and even rights. This is a politics that inserts humans into the earth's logics, with the vegetal as a central and primary force.

A delusional separation of humans from the Earth dictates a narrow view of politics. Plant influences and socialities in this volume appear in diverse forms and open up political possibilities. Robin Wall Kimmerer speaks of Maple Nation in her upstate New York home region and across maple trees' geographic range. In this same way, Denmark could be known as Willow Nation, recognizing the many roles willow trees play in improving the land base. Ecological practices of sharing resources are actually political and are social contracts, French philosopher Michel Serres argues. "In fact, the Earth speaks to us in terms of forces, bonds, and interactions, and that's enough to make a contract" (Serres 1995:39). In an eco or bio-politic, equitable resource distribution would trump inequality, and plant actors become central agents of equity and justice.

New stories as forms of poiesis are needed to co-engender a more-than-human politics. The poem in the ash tree chapter, written by a community concerned about ash tree death, is a prime example. Lisa Jevbratt's weavings, poems to ash trees, are quietly drawing threads together between each and the heart of the world that beats to a pulse we can listen to and for. These new meanings and things are expressions of complexity as intelligence, with each being as a source, foundational to politics and shared abilities to thrive.

References

Abram, David. 2017. "Storytelling and wonder: On the rejuvenation of oral culture as an ecological imperative." In *Ethical Transformations for a Sustainable Future,* edited by O. Urbain and D. Temple, 9–17. Oxon: Routledge.

Abram, David, Tema Milstein, and José Castro-Sotomayor. 2020. "Interbreathing ecocultural identity in the Humilocene." In *Routledge Handbook of Ecocultural Identity,* edited by Tema Milstein and José Sotomayor, 5–25. Oxon: Routledge.

Bacon, J. M. 2019. "Settler colonialism as eco-social structure and the production of colonial ecological violence." *Environmental Sociology* 5 (1): 59–69.

Bakhtin, M. 2014. "Polyphonic discourse in the novel." In *The Discourse Studies Reader: Main Currents in Theory and Analysis,* edited by J. Angermuller, D. Maingueneau, and R. Wodak, 27–35. Amsterdam: John Benjamins Publishing.

Balding, Mung, and Kathryn J. H. Williams. 2016. "Plant blindness and the implications for plant conservation." *Conservation Biology* 30 (6): 1192–1199. https://doi.org/10.1111/cobi.12738.

Bandara, J. M. R. S. 2007. *Nature Farming: Integration of Traditional Knowledge Systems with Modern Farming in Rice.* Leusden: ETC/COMPAS.

Barad, Karen. 2007. *Meeting the Universe Halfway: Quantum Physics and the Entanglement of Matter and Meaning.* Durham: Duke University Press.

Barry, Janice, and Julian Agyeman. 2020. "On belonging and becoming in the settler-colonial city: Co-produced futurities, placemaking, and urban planning in the United States." *Journal of Race, Ethnicity and the City* 1 (1–2): 22–41.

Barua, Maan. 2014a. "Circulating elephants: Unpacking the geographies of a cosmopolitan animal." *Transactions of the Institute of British Geographers* 39 (4): 559–573.

Barua, Maan. 2014b. "Volatile ecologies: Towards a material politics of human—animal relations." *Environment and Planning A* 46 (4): 1462–1478.

Barua, Maan. 2016. "Encounter." *Environmental Humanities* 7 (1): 265–270. https://doi.org/10.1215/22011919-3616479.

Barua, Maan. 2023. "Plantationocene: A vegetal geography." *Annals of the American Association of Geographers* 113 (1): 13–29.

Bateson, Gregory. 1972. *An Ecology of Mind.* New York: Ballentine.

Bateson, Gregory, and Mary Catherine Bateson. 1988. *Angels Fear: Towards an Epistemology of the Sacred.* New York: Bantam Dell Publishing Group.

Begum, Sajida, and I. C. Barua. 2012. "Ficus species and its significance." *International Journal of Computer Applications in Engineering Sciences* 2 (3): 273–275.

Benadusi, Mara. 2015. "Elephants never forget: Capturing nature at the border of Ruhuna National Park (Yala), Sri Lanka." *Capitalism Nature Socialism* 26 (1): 77–96. https://doi.org/10.1080/10455752.2014.971419.

Benedict, Michael, Kelly Kindscher, and Raymond Pierotti. 2014. "Learning from the land: Incorporating indigenous perspectives into the plant sciences." In *Innovative Strategies for Teaching in the Plant Sciences*, edited by Cassandra L. Quave, 135–154. New York: Springer.

Bennett, Jane. 2010. *Vibrant Matter: A Political Ecology of Things.* Durham: Duke University Press.

Berkes, Fikret. 2017. *Sacred Ecology.* New York: Routledge.

Better, Ori. S., Irit Rubinstein, Joseph M. Winaver, and James P. Knochel. 1997. "Mannitol therapy revisited." *Kidney International* 52 (4): 886–894.

Briggs, Katharine M. 1978. *The Vanishing People: A Study of Traditional Fairy Beliefs.* London: Batsford.

Brown, A. D., S. Pacheco, T. Lomáscolo, and L. Malizia. 2006. "Situación ambiental en los bosques andinos yungueños." In *Situación Ambiental Argentina,* edited by A. D Brown, U. Martínez Ortiz, M. Acerbi, and J. Corchera, 53–71. Buenos Aires: Fundación Vida Silvestre Argentina.

Brown, Robert. 1868. "On the vegetable products used by the Northwest American Indians as food and medicine, in the arts and in superstitious rites." *Botanical Society: Edinburgh Transactions* 9: 378–396. Reprinted in 1986, *The North American Indian.*

Brundrett, Mark C., and Leho Tedersoo. 2020. "Resolving the mycorrhizal status of important northern hemisphere trees." *Plant Soil* 454: 3–34.

Campos-Arceiz, Ahimsa, and Steve Blake. 2011. "Megagardeners of the forest – the role of elephants in seed dispersal." *Acta Oecologica* 37 (6): 542–553. https://doi.org/10.1016/j.actao.2011.01.014.

Carr, John, and Tema Milstein. 2021. "See nothing but beauty": The shared work of making anthropogenic destruction invisible to the human eye. *Geoforum* 122 (12): 183–192.

Casey, Edward. 1993. *Getting Back into Place: Toward a Renewed Understanding of Place-World.* Indianapolis: Indiana University Press.

Casey, Edward S., and Michael Marder. 2023. *Plants in Place: A Phenomenology of the Vegetal.* New York: Columbia University Press.

Chartier, Laura, Alexandra Zimmermann, and Richard J. Ladle. 2011. "Habitat loss and human–elephant conflict in Assam, India: Does a critical threshold exist?" *Oryx* 45 (4): 528–533. https://doi.org/10.1017/S0030605311000044.

Childe, V. Gordon. 1928. *The Most Ancient East: The Oriental Prelude to European Prehistory*. London: Kegan Paul, Trench, Trubner.

Collie, Madeleine. 2021. "Ash stories: A spell against forgetting." *Performance Philosophy* 6 (2): 156–173.

Commons, Michael B. 2018. "Forest gardening with space and place for wild elephants" [WWW Document]. *Regeneration International*. Accessed December 14, 2020. https://regenerationinternational.org/2018/01/15/forest-gardening-space -place-wild-elephants/.

Cooper, Daniel G., and Nina Kruglikova. 2022. "Psychogeography reimagined." *Journal of Global Indigeneity* 6 (2): 1–12.

Cooper-White, P. 2014. "Intersubjectivity." In *Encyclopedia of Psychology and Religion*, edited by D. A. Leeming. Boston: Springer. https://doi.org/10.1007 /978-1-4614-6086-2_9182.Costelloe, Justin F., Jane Leeder, and Marcus Strang. 2016. "Drivers of the distribution of a dominant riparian tree species (Eucalyptus coolabah) on a dryland river system, Diamantina River, Australia." In *Proceedings 11th International Symposium on Ecohydraulics* (Vol. 26667), University of Melbourne.

Cresswell, Tim. 1996. *In Place/Out of Place: Geography, Ideology, Transgression*. Minneapolis: University of Minnesota Press.

Crosby, A., and D. Worster. 1988. "Ecological imperialism: The overseas migration of western Europeans as a biological phenomenon." In *American Encounters: Natives and Newcomers from European Contact to Removal—1500–1800*, edited by P. C. Mancall and J. H. Merrell, 55–67. New York: Routledge.

Darby, Melissa Cole. 1996. "Wapato for the people: An ecological approach to understanding the native American use of *Sagittaria latifolia* on the Lower Columbia River." MA thesis. Portland State University.

Davies, William. 2019. *Nervous States: Democracy and the Decline of Reason*. New York: W.W. Norton & Company.

Davis, W. 2001. *Light at the Edge of the World: A Journey through the Realm of Vanishing Cultures*. Vancouver: Douglas and McIntyre.

Debord, Guy. 1956. "Theory of the dérive." *Situationalist International Online*. Retrieved from http://www.cddc.vt.edu/sionline/si/theory.html.

Delijani, Sahar. 2014. *Children of the Jacaranda Tree*. New York: Atria Books.

Denham, Michael A. 1846. *A Collection of Proverbs and Popular Sayings Relating to the Seasons, the Weather and Agricultural Pursuits*. London: Percy Society.

Descola, Philippe. 2005. *Par-delà nature et culture*. Paris: Éditions Gallimard.

De Silva, C. S., E. K. Weatherhead, J. W. Knox, and J. A. Rodriguez-Diaz. 2007. "Predicting the impacts of climate change—A case study of paddy irrigation water requirements in Sri Lanka." *Agricultural Water Management* 93: 19–29. https://doi .org/10.1016/j.agwat.2007.06.003.

De Silva, S., and K. Srinivasan. 2019. "Revisiting social natures: People-elephant conflict and coexistence in Sri Lanka." *Geoforum* 102: 182–190. https://doi.org/10 .1016/j.geoforum.2019.04.004.

Dev, Laura. 2018. "Plant knowledges: Indigenous approaches and interspecies listening toward decolonizing ayahuasca research." In *Plant Medicines, Healing and Psychedelic Science,* edited by B. C. Labate and C. Cavnar, 185–204. Berlin: Springer Nature.

Dreyfus, Hubert, and Sean D. Kelly. 2011. *All Things Shining: Reading the Western Classics to Find Meaning in a Secular Age.* New York: Free Press.

Dubost, Jean-Marc, Vichith Lamxay, Sabrina Krief, Michael Falshaw, Chanthanom Manithip, and Eric Deharo. 2019. "From plant selection by elephants to human and veterinary pharmacopeia of mahouts in Laos." *Journal of Ethnopharmacology* 244 (4): 112157. https://doi.org/10.1016/j.jep.2019.112157.

Elton, Sarah. 2021. "Growing methods: Developing a methodology for identifying plant agency and vegetal politics in the city." *Environmental Humanities* 13 (1): 93–112.

Escobar, Arturo. 1998. "Whose knowledge, Whose nature? biodiversity, conservation, and the political ecology of social movements." *Journal of Political Ecology* 5: 53–82.

Escobar, Arturo. 2001. "Culture sits in places: Reflections on globalism and subaltern strategies of localization." *Political Geography* 20 (2): 139–174.

Esham, Mohamed, Brent Jacobs, Hewage Sunith Rohitha Rosairo, and Balde Boubacar Siddighi. 2017. "Climate change and food security: A Sri Lankan perspective." *Environment, Development, and Sustainability* 20: 1017–1036. https://doi.org/10.1007/s10668-017-9945-5.

Evans, Lauren A., and William A. Adams. 2018. "Elephants as actors in the political ecology of human-elephant conflict." *Transactions of the Institute of British Geographers* 43: 630–645. https://doi.org/10.1111/tran.12242.

Fernando, Prithiviraj. 2000. "Elephants in Sri Lanka: Past, present and future." *Loris* 22: 38–44.

Fernando, Prithiviraj. 2015a. "The starving elephants of Udawalawe." *Sanctuary Asia* 35.

Fernando, Prithiviraj. 2015b. "Managing elephants in Sri Lanka: Where we are and where we need to be." *Ceylon Journal of Science* 44: 1–11.

Fernando, Prithiviraj, Eric Wikramanayake, Devaka Weerakoon, L. K. A. Jayasinghe, Manori Gunawardene, and H. K. Janaka. 2005. "Perceptions and patterns of human–elephant conflict in old and new settlements in Sri Lanka: Insights for mitigation and management." *Biodiversity and Conservation* 14 (10): 2465–2481. https://doi.org/10.1007/s10531-004-0216-z.

Fernando, Prithiviraj, and Peter Leimgruber. 2011. "Asian elephants and seasonally dry forests." In *The Ecology and Conservation of Seasonally Dry Forests in Asia,* edited by W. J. McShea, S. J. Davies, N. Phumpakphan, and A. Pattanavibool, 151–163. Washington, DC: Smithsonian Institution Scholarly Press.

Fleming, Jake. 2017. "Toward vegetal political ecology: Kyrgyzstan's walnut–fruit forest and the politics of graftability." *Geoforum* 79: 26–35.

Forester, John F. 2021. *How Spaces become Places: Place Makers Tell Their Stories.* New York: New Village Press.

Foster, John Bellamy. 1999. "Marx's theory of metabolic rift: Classical foundations for environmental sociology." *American Journal of Sociology* 105: 366–405.

Frawley, Jodi. 2010. "Detouring to grafton: The Sydney botanic gardens and the making of an Australian urban aesthetic." *Australian Humanities Review* 49: 119–139.

Gadgil, Madhav, and Ramachandra Guha. 2013. *Ecology and Equity: The Use and Abuse of Nature in Contemporary India*. London: Routledge.

Gagliano, Monica. 2015. "In a green frame of mind: Perspectives on the behavioural ecology and cognitive nature of plants." *AoB PLANTS* 7. https://doi.org/10.1093/aobpla/plu075.

Gagliano, Monica. 2018. *Thus Spoke the Plant: A Remarkable Journey of Groundbreaking Scientific Discoveries and Personal Encounters with Plants*. Berkeley: North Atlantic Books.

Gagliano, Monica, Michael Renton, Martial Depczynski, and Stefano Mancuso. 2014. "Experience teaches plants to learn faster and forget slower in environments where it matters." *Oecologia* 175: 63–72. https://doi.org/10.1007/s00442-013-2873-7.

Gagliano, Monica, John C. Ryan, and Patrícia Vieira. 2017. "Introduction." In *The Language of Plants: Science, Philosophy, Literature*, edited by Monica Gagliano, John C. Ryan, and Patrícia Vieira, vii–xxxiii. Minneapolis: University of Minnesota Press.

Garber, M. "The dark side of the houseplant boom." *The Atlantic*. Accessed April 20, 2021. www. theatlantic.com/culture/archive/2021/04/dark-side-houseplant-boom-nature-empathy/618638/.

Garibaldi, Ann Catherine. 2003. "Bridging ethnobotany, autecology, and restoration: The study of Wapato *Sagittaria latifolia Willd; Alismataceae* in interior British Columbia." Masters of Science thesis. University of Victoria.

Garibaldi, A., and Nancy Turner. 2004. "Cultural keystone species: Implications for ecological conservation and restoration." *Ecology and Society* 9 (3): 1. http://www.ecologyandsociety.org/vol9/iss3/art1/.

Ghosh, Amitav. 2021. *The Nutmeg's Curse*. Chicago: University of Chicago Press.

Gianinazzi, Silvio, Armelle Gollotte, Marie-Noelle Binet, Diederik van Tuinen, Dirk Redecker, and Daniel Wipf. 2010. "Agroecology: The key role of arbuscular mycorrhizas in ecosystem services." *Mycorrhiza* 20: 519–530. https://doi.org/10.1007/s00572-010-0333-3.

Gorzelak, Monika A., Amanda K. Asay, Brian J. Pickles, and Suzanne W. Simard. 2015. "Inter-plant communication through mycorrhizal networks mediates complex adaptive behaviour in plant communities." *AoB PLANTS* 7: plv050. https://doi.org/10.1093/aobpla/plv050.

Graeber, David, and David Wengrow. 2021. *The Dawn of Everything: A New History of Humanity*. London: Penguin UK.

Grosz, Elizabeth. 2004. *The Nick of Time: Politics, Evolution, and the Untimely*. Durham: Duke University Press.

Gu, Chuanhui, John Crane II, George Hornberger, and Amanda Carrico. 2015. "The effects of household management practices on the global warming potential of urban lawns." *Journal of Environmental Management* 151: 233–242.

Guattari, Félix. 2005. *The Three Ecologies*. London: Bloomsbury Publishing.

Guichard, P., J. P. Peltier, E. Gout, R. Bligny, and G. Marigo. 1997. "Osmotic adjustment in Fraxinus excelsior L.: malate and mannitol accumulation in leaves under drought conditions." *Trees* 11 (3): 155–161.

Gunasena, H. P. M., and D. K. N. G. Pushpakumara. 2015. "Chena cultivation in Sri Lanka: Prospects for agroforestry interventions." In *Shifting Cultivation and Environmental Change: Indigenous People, Agriculture and Forest Conservation,* edited by M. Cairns, 199–200. New York: Routledge.

Gutsche Jr., Robert E. 2014. "News place-making: Applying 'mental mapping' to explore the journalistic interpretive community." *Visual Communication* 13 (4): 487–510.

Gutsche, Jr, Robert E., and Kristy Hess. 2020. "Placeification: The transformation of digital news spaces into 'places' of meaning." *Digital Journalism* 8 (5): 586–595.

Hagerhall, Caroline M., Thorbjörn Laike, Richard P. Taylor, Marianne Küller, Rikard Küller, and Theodore P. Martin. 2008. "Investigations of human EEG response to viewing fractal patterns." *Perception* 37 (10): 1488–1494.

Hajda, Yvonne P. 1984. "Regional social organization in the greater lower Columbia, 1792–1830." PhD thesis. University of Washington.

Hall, Matthew. 2011. *Plants as Persons: A Philosophical Botany.* Albany: State University of New York Press.

Hall, Matthew. 2019a. *The Imagination of Plants: A Book of Botanical Mythology.* Albany: State University of New York Press.

Hall, Matthew. 2019b. "In defense of plant personhood." *Religions* 10 (5): 317.

Hallowell, Alfred Irving. 1960. "Ojibwa ontology, behavior and world view." In *Culture in History: Essays in Honor of Paul Radin,* edited by Stanley Diamond, 19–52. New York: Octagon Books.

Hamdi, Nabeel. 2010. *The Placemaker's Guide to Building Community.* London: Earthscan.

Handawela, James. 2016. *Ancient Dry Zone Watershed Farming System: Hena, Wewa, and Purana Evolved by Farmers and Failed by Rulers.* Colombo: S. Godage and Brothers Ltd.

Haraway, Donna. 2003. *The Companion Species Manifesto: Dogs, People, and Significant Otherness.* Vol. 1. Chicago: Prickly Paradigm Press.

Haraway, Donna. 2008. *When Species Meet, Posthumanities.* Vol. 3. Minneapolis: University of Minnesota Press.

Haraway, Donna. 2015. "Anthropocene, Capitalocene, Plantationocene, Chthulucene: Making kin." *Environmental Humanities* 6 (1): 159–165.

Harries-Jones, Peter. 1995. *A Recursive Vision: Ecological Understanding and Gregory Bateson.* Toronto: University of Toronto Press.

Harries-Jones, Peter. 2005. "Gregory Bateson and ecological aesthetics: An introduction." *Australian Humanities Review* 35.

Harrison, Robert Pogue. 1992. *Forests: The Shadow of Our Civilization.* Chicago: University of Chicago Press.

Hartigan Jr, John. 2017. *Care of the Species: Races of Corn and the Science of Plant Biodiversity.* Minneapolis: University of Minnesota Press.

Hartigan Jr, John. 2019. "Plants as ethnographic subjects." *Anthropology Today* 35 (2): 1–2.

Hawken, Paul. 1993. *The Ecology of Commerce: A Declaration of Sustainability.* New York: Harper Business.

Head, Lesley, and Jennifer Atchison. 2009. "Cultural ecology: emerging human-plant geographies." *Progress in Human Geography* 33 (2): 236–245.

Head, Lesley M., and Pat Muir. 2005. "Living with trees—Perspectives from the suburbs." In *Proceedings of the 6th National Conference of the Australian Forest History Society* edited by M. Calver, H. Bigler-Cole, G. Bolton, Gaynor A. Dargavel, P. Horwitz, J. Mills, and G. Wardell-Johnson, 84–95. Rotterdam: Millpress.

Head, Leslie, Jennifer Atchison, Catherine Phillips, and Kathleen Buckingham. 2014. "Vegetal politics: Belonging, practices and places." *Social & Cultural Geography* 15 (8): 861–870. https://doi.org/10.1080/14649365.2014.973900.

Henle, Klaus, Didier Alard, Jeremy Clitherow, Paul Cobb, Les Firbank, Tiiu Kull, Davy McCracken, Robin F. A. Moritz, Jari Niemelä, Michael Rebane, Dirk Wascher, Allan Watt, and Juliette Young. 2008. "Identifying and managing the conflicts between agriculture and biodiversity conservation in Europe—A review." *Agriculture, Ecosystems & Environment, Special Section: Problems and Prospects of Grassland Agroecosystems in Western China* 124: 60–71. https://doi.org/10.1016/j.agee.2007.09.005.

Hills, R. 2020. "Jacaranda mimosifolia." The IUCN Red List of Threatened Species: e.T32027A68135641. Accessed July 03, 2023. https://doi.org/10.2305/IUCN.UK.2020-3.RLTS.T32027A68135641.en.

Hinsinger, D. D., J. Basak, M. Gaudeul, C. Cruaud, P. Bertolino, N. Frascaria-Lacoste, and J. Bousquet. "The phylogeny and biogeographic history of ashes (fraxinus, oleaceae) highlight the roles of migration and vicariance in the diversification of temperate trees." *PLoS One*. Accessed November 21, 2013; 8 (11): e80431. https://doi.org/10.1371/journal.pone.0080431.

Hoare, Richard E., and Johan T. Du Toit. 1999. "Coexistence between people and elephants in African Savannas." *Conservation Biology* 13 (3) 633–639. https://doi.org/10.1046/j.1523-1739.1999.98035.x.

Huang, Zhongwei, Mohamad Hejazi, Qiuhong Tang, Chris R. Vernon, Yaling Liu, Min Chen, and Kate Calvin. 2019. "Global agricultural green and blue water consumption under future climate and land use changes." *Journal of Hydrology* 574: 242–256.

Humboldt, Alexander von. 2016. *Views of Nature*. Edited by Stephen T. Jackson and Laura Dassow Walls. Chicago: University of Chicago Press.

Ingold, Tim. 1999. "Three in one: On dissolving the distinction between body, mind, and culture." Unpublished manuscript. University of Manchester.

Ingold, Tim. 2000. *Perception of the Environment: Essays on Livelihood, Dwelling, and Skill*. London: Routledge.

Ingold, Tim. 2003. "Two reflections on ecological knowledge." In *Nature Knowledge: Ethnoscience, Cognition, and Utility*, edited by Glauco Sanga and Gherardo Ortalli, 301–311. Oxford: Berghahn Books.

Ingold, Tim. 2017. "On human correspondence." *Journal of the Royal Anthropological Institute* 23 (1): 9–27.

Ingold, Tim. 2021. *Correspondences*. Cambridge: Polity.

Isthikar, M. A. M. 2015. "Seasonal feeding ecology of the elephants in the Udawalawe National Park, Sri Lanka: (A Geographical Survey)." *Journal of Social Sciences* 7 (4): 213–218.

Jiaen, Zhang, Quan Guoming, Zhao Benliang, Liang Kaiming, and Qin Zhong. 2017. "Rice–Duck co-culture in China and its ecological relationships and functions." In *Agroecology in China: Science, Practice, and Sustainable Management,* edited by Luo Shiming and Stephen R. Gliessman, 111–138. Boca Raton: CRC Press.

Kamau, Peter N., and Andrew Sluyter. 2018. "Challenges of elephant conservation: Insights from oral histories of colonialism and landscape in Tsavo, Kenya." *Geographical Review* 108 (4): 523–544. https://doi.org/10.1111/gere.12288.

Kane, Paul. 1859. *Wanderings of an Artist Among the Indians of North America, From Canada to Vancouver's Island and Oregon through the Hudson's Bay Company's Territory and Back Again.* London: Langmans, Brown, Green, Longmans, and Roberts.

Karp, Daniel, Andrew J. Rominger, Jim Zook, Jai Ranganathan, Paul R. Ehrlich, and Gretchen C. Daily. 2012. "Intensive agriculture erodes β-diversity at large scales." *Ecology Letters* 15 (9): 963–970. https://doi.org/10.1111/j.1461-0248.2012.01815 .x.

Khan, Gulshan Ara. 2012. "Vital materiality and non-human agency: An interview with Jane Bennett." In *Dialogues with Contemporary Political Theorists,* edited by Gary Browning, Raia Prokhovnik, and Maria Dimova-Cookson, 42–57. London: Palgrave Macmillan UK.

Kimmerer, Robin Wall. 2003. *Gathering Moss: A Natural and Cultural History of Mosses.* Corvalis: Oregon State University Press.

Kimmerer, Robin Wall. 2015. *Braiding Sweetgrass.* Minneapolis: Milkweed Editions.

King, Lucy E., Ian Douglas-Hamilton, and Fritz Vollrath. 2007. "African elephants run from the sound of disturbed bees." *Current Biology* 17 (19): R832–R833. https://doi.org/10.1016/j.cub.2007.07.038.

King, Lucy E., Ian Douglas-Hamilton, and Fritz Vollrath. 2011. "Beehive fences as effective deterrents for crop-raiding elephants: Field trials in northern Kenya." *African Journal of Ecology* 49: 431–439. https://doi.org/10.1111/j.1365-2028.2011 .01275.x.

King, Lucy, Michael Pardo, Sameera Weerathunga, T. V. Kumara, Nilmini Jayasena, Joseph Soltis, and Shermin de Silva. 2018. "Wild Sri Lankan elephants retreat from the sound of disturbed Asian honey bees." *Current Biology* 28 (2): R64–R65. https://doi.org/10.1016/j.cub.2017.12.018.

Knudsen, Britta Timm, and Carsten Stage. 2015. "Introduction: Affective methodologies." In *Affective Methodologies: Developing Cultural Research Strategies for the Study of Affect,* edited by B. T. Knudsen and C. Stage, 1–22. London: Palgrave Macmillan UK.

Kohák, Erazim. 1984. *The Embers and the Stars: A Philosophical Inquiry into the Moral Sense of Nature.* Chicago: University of Chicago Press.

Kremen, Claire, 2015. "Reframing the land-sparing/land-sharing debate for biodiversity conservation." *Annals of the New York Academy of Sciences* 1355: 52–76. https://doi.org/10.1111/nyas.12845.

Kumar, M. Ananda, Divya Mudappa, and T. R. Shankar Raman. 2010. "Asian Elephant (Elephas Maximus) habitat use and ranging in fragmented Rainforest and plantations in the Anamalai Hills, India." *Tropical Conservation Science* 3 (2): 143–158. https://doi.org/10.1177/194008291000300203.

Latour, Bruno. 2005. *Reassembling the Social: An Introduction to Actor-Network-Theory.* Oxford: Oxford University Press.

Latour, Bruno. 2018. *Down to Earth: Politics in the New Climatic Regime.* Cambridge: Polity Press.

Lawrence, Anna M. 2022. "Listening to plants: Conversations between critical plant studies and vegetal geography." *Progress in Human Geography* 46 (2): 629–651.

Lawrence-Zuniga, Denise. 2017. "Space and place." *Oxford Bibliographies.* Oxford: Oxford University Press.

Leimgruber, P., J. B. Gagnon, C. Wemmer, D. S. Kelly, M. A. Songer, and E. R. Selig. 2003. "Fragmentation of Asia's remaining wildlands: Implications for Asian elephant conservation." *Animal Conservation* 6: 347–359. https://doi.org/10.1017/S1367943003003421.

Lien, Marianne Elisabeth, Heather Anne Swanson, and Gro Ween. 2018. "Introduction." In *Domestication Gone Wild: Politics and Practices of Multispecies Relations,* 1–32. Durham: Duke University Press.

Lindbladh, Johanna. 2017. "The polyphonic performance of testimony in Svetlana Aleksievich's Voices from Utopia." *Canadian Slavonic Papers* 59 (3–4): 281–312. https://doi.org/10.1080/00085006.2017.1379116.

Lomolino, Mark V. 2020. *Biogeography: A Very Short Introduction.* Oxford: Oxford University Press.

Lopez, Barry. 1997. *A Literature of Place.* Portland: Portland Magazine.

Luoma, Jon R. 1999. *The Hidden Forest: The Biography of an Ecosystem.* New York: Henry Holt Publishing.

Lyons, Natasha, Tanja Hoffmann, Debbie Miller, Stephanie Huddlestan, Roma Leon, and Kelly Squires. 2018. "Katzie & the Wapato: An archaeological love story." *Archaeologies* 14 (1): 7–29.

Mabey, Richard. 1996. *Flora Britannica.* London: Sinclair Stevenson.

Mac Coitir, Niall. 2015. *Ireland's Trees. Myths, Legends and Folklore.* Cork: The Collins Press.

Mancuso, Stefano. 2018. *The Revolutionary Genius of Plants.* New York: Simon and Schuster.

Marder, Michael. 2012. "Resist like a plant! On the vegetal life of political movements." *Peace Studies Journal* 5: 24–32.

Marder, Michael. 2013. *Plant-Thinking: A Philosophy of Vegetal Life.* New York: Columbia University Press.

Marder, Michael. 2016. *Through Vegetal Being: Two Philosophical Perspectives.* New York: Columbia University Press.

Margulies, Jared. 2023. *The Cactus Hunters: Desire and Extinction in the Illicit Succulent Trade.* Minneapolis: University of Minnesota Press.

Margulis, Lynn, and Dorian Sagan. 2023. *Gaia and Philosophy.* London: Ignota.

Marx, Karl. 1976. *Capital.* London: Penguin.

Marx, Karl. 2005. *Grundrisse: Foundations of the Critique of Political Economy.* London: Penguin.

Massey, Doreen. 1991. "A global sense of place." *Marxism Today* 38: 24–29.

Matson, Pamela A., William J. Parton, Alison G. Power, and Michael J. Swift. 1997. "Agricultural intensification and ecosystem properties." *Science* 277 (5325): 504–509. https://doi.org/10.1126/science.277.5325.504.

McIntosh, Alistair. 2004. *Soil and Soul: People Versus Corporate Power.* Folkstone: Arum.

Memory Paterson, Jacqueline. 1997. *Tree Wisdom: The Definitive Guidebook to the Myth, Folklore and Healing Power of Trees.* New York: Harper Collins Publishers.

Mentore, Laura. 2012. "The intersubjective life of cassava among the Waiwai." *Anthropology and Humanism* 37 (2): 146–155.

Merleau-Ponty, Maurice. 2012. *Phenomenology of perception.* Vol. 26. Translated by D. A. Landes. Abingdon, Oxon: Routledge.

Midgley, David. 2004. *The Essential Mary Midgley.* Abingdon, Oxon: Routledge.

Milstein, Tema, and José Castro-Sotomayor. 2020. "Ecocultural identity." In *Routledge Handbook of Ecocultural Identity,* edited by Tema Milstein, and José Castro-Sotomayor. Oxon: Routledge.

Mishra, Stuti. 2020. *The independent.* Accessed December 15, 2020. https://www.independent.co.uk/news/world/asia/sri-lanka-elephants-rubbish-landfill-b1761667.html.

Moran, Robyn, and Lisbeth A. Berbary. 2021. "Placemaking as unmaking: Settler colonialism, gentrification, and the myth of 'revitalized' urban spaces." *Leisure Sciences* 43 (6): 644–660.

Morphy, Howard. 1989. "From dull to brilliant: The aesthetics of spiritual power among the Yolngu." *Man* 24 (1): 21–40.

Mostafa, Nada M., Omayma A. Eldahshan, and Abdel Nasser B. Singab. 2014. "The genus Jacaranda (Bignoniaceae): An updated review." *Pharmacognosy Communications* 4 (3): 31–39.

Muratet, Audrey, Patricia Pellegrini, Anne-Béatrice Dufour, Teddy Arrif, and François Chiron. 2015. "Perception and knowledge of plant diversity among urban park users." *Landscape and Urban Planning* 137: 95–106.

Myers, Natasha. 2017. "From the anthropocene to the planthroposcene: Designing gardens for plant/people involution." *History and Anthropology* 28 (3): 297–301.

Myers, Natasha. 2020. "Becoming sensor in sentient worlds: A more-than-natural history of a black oak savannah." In *Between Matter and Method* edited by Gretchen Bakke and Marina Peterson, 73–96. London: Routledge.

Nature Conservancy. 2021. *Tribal Forest Management.* https://www.nature.org/en-us/about-us/where-we-work/united-states/minnesota/stories-in-minnesota/nature-climate-solutions/tribal-forest-management/.

Nelson, Gerald C., Mark W. Rosegrant, Jawoo Koo, Richard Robertson, Timothy Sulser, Tingju Zhu, Claudia Ringler, Siwa Msangi, Amanda Palazzo, Miroslav Batka, Marilia Magalhaes, Rowena Valmonte-Santos, Mandy Ewing, and David Lee. 2009. Food Policy Report: "Climate Change: Impact on Agriculture and Costs of Adaptation." Washington, DC: International Food Policy Research Institute.

Nyhus, Philip, and Ronald Tilson. 2004. "Agroforestry, elephants, and tigers: Balancing conservation theory and practice in human-dominated landscapes of Southeast

Asia." *Agriculture, Ecosystems & Environment* 104: 87–97. https://doi.org/10.1016/j.agee.2004.01.009.

O'Connor, Jim E., Victor R. Baker, Richard B. Waitt, Larry N. Smith, Charles M. Cannon, David L. George, and Roger P. Denlinger. 2020. "The Missoula and Bonneville floods—A review of ice-age megafloods in the Columbia River basin." *Earth-Science Reviews* 208: 103181.

Oele, Marjolein. 2020. *E-Co-Affectivity: Exploring Pathos at Life's Material Interfaces.* Albany: State University of New York Press.

Oriel, Elizabeth. 2022. "A field guide to human-elephant relations in Sri Lanka: Patterns, roles, and rhythms of multispecies socialities within conflict and cohabitation." PhD Thesis. University of London.

Oriel, Elizabeth, and Toni Frohoff. 2020. "Interspecies ecocultural identities in human-elephant cohabitation." In *The Routledge Handbook of Ecocultural Identities,* edited by Tema Milstein and Jose Castro-Sotomayor. London: Routledge.

Parker, Edward. 2021. *Ash.* London: Reaktion Books.

Pedersen, Stein, Jakob Valentin, Bruno Latour, and Nikolaj Schultz. 2019. "A conversation with Bruno Latour and Nikolaj Schultz: Reassembling the geo-social." *Theory, Culture & Society* 36 (7–8): 215–230.

Phalan, Ben, Malvika Onial, Andrew Balmford, and Rhys E. Green. 2011. "Reconciling food production and biodiversity conservation: Land sharing and land sparing compared." *Science* 333 (6047): 1289–1291. https://doi.org/10.1126/science.1208742.

Phillips, Catherine, and Eily Schulz. 2021. "Greening home: Caring for plants indoors." *Australian Geographer* 52 (4): 373–389.

Plumwood, Val. 2002. *Environmental Culture: The Ecological Crisis of Reason.* Oxon: Routledge.

Plumwood, Val. 2005. "Decolonising Australian gardens: Gardening and the ethics of place." *Australian Humanities Review* 36: 1–9.

Plumwood, Val. 2013. "Nature in the active voice." In *The Handbook of Contemporary Animism,* edited by Graham Harvey, 441–453. Durham: Acumen.

Pollan, Michael. 2001. *The Botany of Desire: A Plant's-Eye View of the World.* New York: Random House.

Prakash, T. G. Supan Lahiru, A. W. Wijernatne, and Prithiviraj Fernando. 2020. "Human-elephant conflict in Sri Lanka: Patterns and extent." *Gajah: Journal of the Asian Elephant Specialist Group* 51: 16–25.

Puig De La Bellacasa, Maria. 2015. "Making time for soil: Technoscientific futurity and the pace of care." *Social Studies of Science* 45 (5): 691–716.

Raffles, Hugh. 2003. "Intimate knowledge." In *Knowledge: Critical Concepts,* edited by Nico Stehr and Reiner Grundmann, 385–398. London: Routledge.

Ranagalage, Manjula, M. H. J. P. Gunarathna, Thilina Surasinghe, D. M. S. L. B. Dissanayake, Matamyo Simwanda, Yuji Murayama, Takehiro Morimoto, Darius Phiri, Vincent R. Nyirenda, Kachchakaduge T. Premakantha, and Anura Sathurusinghe. 2020. "Multi-decadal forest-cover dynamics in the tropical realm: Past trends and policy insights for forest conservation in dry zone of Sri Lanka." *Forests* 11 (8): 836. https://doi.org/10.3390/f11080836.

Ratnapala, Nandasena. 1980. *Sinhalese Folk Lore, Folk Religion, and Folk Life*. Sarvodaya Publishers.

Reuther, B. 2014. "Intersubjectivity, overview." In *Encyclopedia of Critical Psychology*, edited by T. Teo, 1001–1005. New York: Springer. https://doi.org/10.1007/978-1-4614-5583-7_459.

Ribó, Ignasi. 2022. "Poetics of cohabitation: An ecosemiotic theory of oral poiesis." *Poetics Today: International Journal for Theory and Analysis of Literature and Communication* 43 (3): 549–581.

Ruphra, Emiliano. 2020. "Where is manna from?" *Atlas Obscura*. Accessed June 25, 2022. Available at: https://www.atlasobscura.com/articles/where-is-manna-from.

Ruppel, S. 2020. "Houseplants and the invention of indoor gardening." In *The Routledge History of the Domestic Sphere in Europe*, edited by J. Eibach and M. Lanzinger, 509–523. London: Routledge.

Rose, Deborah, Bird. 1999. "Indigenous ecologies and an ethic of connection." In *Global Ethics and Environment*, edited by Nicholas Low, 175–187. London: Routledge.

Rose, Deborah Bird. 2002. "Dialogue with place: Toward an ecological body." *Journal of Narrative Theory* 32 (3): 311–325.

Rose, Deborah Bird. 2017. "Shimmer: When all you love is being trashed." In *Arts of Living on a Damaged Planet*, edited by Anna L. Tsing, Nils Bubandt, Elaine Gan, and Heather Swanson. Minneapolis: University of Minnesota Press.

Rose, Deborah Bird. 2022. *Shimmer: Flying Fox Exuberance in Worlds of Peril*. Edinburgh: Edinburgh University Press.

Rose, Deborah Bird, Thom Van Dooren, and Matthew Chrulew. 2017. "Introduction." In *Extinction Studies: Stories of Time, Death, and Generations*, edited by Deborah Bird Rose, Thom Van Dooren, and Matthew Chrulew, 1–19. New York: Columbia University Press.

Ryan, John C. 2022. "'Dressed in native trees': Plants as figures of anti-national resistance in contemporary aboriginal Australian poetry." In *Global Perspectives on Nationalism: Political and Literary Discourses*, edited by Debajyoti Biswas, Panos Eliopoulos, and John Ryan, 243–259. Oxon: Routledge.

Ryan, John C., Patrícia I. Vieira, and Monica Gagliano. 2021. *The Mind of Plants: Narratives of Vegetal Intelligence*. Santa Fe: Synergetic Press.

Sagan, Dorion. 2023. "Introduction." In Dorian Sagan, and Lynn Margulis (writers) *Gaia and Philosophy*. London: Ignota.org.

Sahlins, Marshall. 2018. "On the ontological scheme of beyond nature and culture." In *Rethinking Relations and Animism: Personhood and Materiality*, edited by Miguel Astor-Aguilera and Graham Harvey, 15–24. London: Routledge.

Salmon, Enrique. 2021. *Iwigara: American Indian Ethnobotanical Traditions and Science*. Portland: Timber Press.

Sanchez, Alberto Ruy. 2019. *Dicen las Jacarandas*. Ciudad de México: Alacena bolsillo.

Santiapillai, Charles, Prithiviraj Fernando, and Manori Gunewardene. 2006. "A strategy for the conservation of the Asian elephant in Sri Lanka." *Gajah* 25: 91–102.

Saville, Dara, and Jessie Wolf Hardin. 2021. *The Ecology of Herbal Medicine.* Albuquerque: University of New Mexico Press.

Schmitt, Melissa H., Adam Shuttleworth, David Ward, and Adrian M. Shrader. 2018. "African elephants use plant odours to make foraging decisions across multiple spatial scales." *Animal Behaviour* 141: 17–27. https://doi.org/10.1016/j.anbehav .2018.04.016.

Scott, James C. 2017. *Against the Grain: A Deep History of the Earliest States.* New Haven: Yale University Press.

Scott, James. 2020. "Facing the anthropocene luce Lecture." *Duke University.* https:// www.youtube.com/watch?v=IwMQSOdLULI.

Scott, James C. 2020. "In praise of floods- Luce Lecture." Accessed December 15, 2023. Available: https://www.youtube.com/watch?v=IwMQSOdLULI.

Seigworth, G. J., and Melissa Gregg. 2010. "An inventory of shimmers." In *The Affect Theory Reader,* edited by Greg J. Seigworth and Melissa Gregg, 1–25. Durham: Duke University Press.

Senanayake, Mahesh. 2018. "Development, politics and disaster mitigation: A case study of the Moragahakanda irrigation project." *Procedia Engineering,* 7th International Conference on Building Resilience: Using scientific knowledge to inform policy and practice in disaster risk reduction 212: 1062–1067. https://doi.org/10 .1016/j.proeng.2018.01.137.

Serres, Michel. 1995. *The Natural Contract.* Ann Arbor: University of Michigan Press.

Shapiro, N., and E. Kirksey. 2017. "Chemo-ethnography: An introduction." *Cultural Anthropology* 32 (4): 481–493.

Shelton, Jo-Anne. 2004. "Killing animals that don't fit in: Moral dimensions of habitat restoration." *Between the Species* 13 (4): 3.

Shiva, Vandana. 1993. *Monocultures of the Mind: Perspectives on Biodiversity and Biotechnology.* New York: Palgrave Macmillan.

Shiva, Vandana. 2016. *The Violence of the Green Revolution: Third World Agriculture, Ecology, and Politics.* Lexington: University Press of Kentucky.

Shurkin, Joel. 2014. "Animals that self-medicate." *Proceedings of the National Academy of Sciences of the United States of America* 111 (49): 17339–17341.

Simard, S. W. 2018. "Mycorrhizal networks facilitate tree communication, learning, and memory." In *Memory and Learning in Plants,* edited by M. Baluska, M. Gagliano, and G. Witzany, 191–213. Springer International Publishing. https://doi .org/10.1007/978-3-319-75596-0_10.

Simard, Suzanne. 2021. *Finding the Mother Tree: Uncovering the Wisdom and Intelligence of the Forest.* London: Penguin.

Simpson, George. 1931. "Fur trade and empire: George Simpson's journal entitled remarks connected with the fur trade in the course of a voyage from York Factory to Fort George and back to York Factory 1824–1825, with related documents." Cambridge: Harvard University Press.

Singh, Neera M. 2018. "Introduction: Affective ecologies and conservation." *Conservation and Society* 16 (1): 1–7.

Skelton, Robert P., Leander D. L. Anderegg, Jessica Diaz, Matthew M. Kling, Prahlad Papper, Laurent J. Lamarque, Sylvain Delzon, Todd E. Dawson, and David

D. Ackerly. 2021. "Evolutionary relationships between drought-related traits and climate shape large hydraulic safety margins in western North American oaks." *Proceedings of the National Academy of Sciences* 118 (10): e2008987118.

Smith, Mick. 2007. "On 'being' moved by nature: Geography, emotion and environmental ethics." In *Emotional Geographies*, edited by Joyce Davidson, Mick Smith, and Liz Bondi, 219–230. Hampshire: Ashgate Publishers.

Spicer, E. H. 1980. *The Yaquis: A Cultural History.* Tuscon: University of Arizona Press.

Stein Pedersen, Jakob Valentin, Bruno Latour, and Nikolaj Schulz. 2019. "A conversation with Bruno Latour and Nikolaj Schultz: Reassembling the geo-social." *Theory, Culture, & Society* 36 (7–8): 215–230.

Stevens, Michelle, Dale C. Darris, and Scott M. Lambert. 2001. "Ethnobotany, culture, management, and use of common camas." *Native Plants Journal* 2 (1): 47–53.

Sukumar, Rajan. 2006. "A brief review of the status, distribution and biology of wild Asian elephants Elephas maximus." *International Zoo Yearbook* 40 (1): 1–8.

Sukumar, Rajan. 2011. *The Story of Asia's Elephants.* Mumba: The Marg Foundation.

Taylor, R. A. J., Leah Bauer, Terese Poland, and Keith Windell. 2010. "Flight performance of Agrilus planipennis (Coleoptera: Buprestidae) on a flight mill and in free flight." *Journal of Insect Behavior* 23: 128–148.

Thakur, M. L., M. Ahmad, and R. K. Thakur. 1992. "Lantana weed (Lantana camara var. aculeata Linn) and its possible management through natural insect pests in India." *Indian Forester* 118: 466–488.

Thakur, A. K., D. K. Yadav, and M. K. Jhariya. 2016. "Socio-economic status of human-elephant conflict: Its assessment and solutions." *Journal of Applied and Natural Science* 8 (4): 2104–2110.

Thompson, Claire. 2022. "Native tribes are bringing native prairie back to the Pacific Northwest." *Civil Eats.* September 16, 2022. https://civileats.com/2022/09/16/native-tribes-prairie-land-ecosystem-restoration-camas-indigenous-foodways-pacific-northwest/.

Torrico, Gualberto, Luis Rea, and Stephan Beck. 1997. *Estudio sobre los árboles y arbustos de uso múltiple en los departamentos de Cochabamba y Chuquisaca (Valles secos interandinos).* La Paz: Instituto de Ecología.

Trewavas, Anthony. 2015. *Plant Behaviour and Intelligence.* Oxford: Oxford University Press.

Trewavas, Anthony. 2016a. "Intelligence, cognition, and language of green plants." *Frontiers in Psychology* 7: 194696. https://doi.org/10.3389/fpsyg.2016.00588.

Trewavas, Anthony. 2016b. "Plant intelligence: An overview." *BioScience* 66 (7): 542–551.

Tsing, Anna L. 2012. "On nonscalability: The living world is not amenable to precision-nested scales." *Common Knowledge* 25 (1–3): 143–162. https://doi.org/10.1215/0961754X-1630424.

Tsing, Anna L. 2013. "More than human sociality." *Anthropology and Nature* 14 (1): 27–42.

Tsing, Anna L. 2015. *The Mushroom at the End of the World: On the Possibility of Life in Capitalist Ruins.* Princeton: Princeton University Press.

Tsing, Anna L., and Jesse Bazzul. 2022. A feral atlas for the Anthropocene: An interview with Anna L. Tsing. In *Reimagining Science Education in the Anthropocene*, edited by Maria F. G. Wallace, Jesse Bazzul, Marc Higgins, and Sara Tolbert, 309–319. London: Palgrave Macmillan.

Turner, Nancy J., Iain J. Davidson-Hunt, and Michael O'Flaherty. 2003. "Living on the edge: Ecological and cultural edges as sources of diversity for social—ecological resilience." *Human Ecology* 31: 439–461.

Ulrich, Roger S. 1984. "View through a window may influence recovery from surgery." *Science* 224 (4647): 420–421.

Van Daele, Wim. 2008. "The meaning of culture-specific food: Rice and its web of significance in Sri Lanka." *Journal of Applied Anthropology*: 292–304.

Van Dooren, Thom. 2017. "Spectral crows in Hawaii: Conservation and the work of inheritance." In *Extinction Studies: Stories of Time, Death, and Generations*, edited by Deborah Bird Rose, Thom Van Dooren, and Matthew Chrulew, 187–216. New York: Columbia University Press.

Von Goethe, Johann Wolfgang, and Gordon L. Miller. 2009. *The Metamorphosis of Plants*. Cambridge: MIT Press.

Wandersee, James H., and Elisabeth E. Schussler. 1999. "Preventing plant blindness." *The American Biology Teacher* 61 (2): 82–86.

Warner, Helen, and Sanna Inthorn. 2022. "Activism to make and do: The (quiet) politics of textile community groups." *International Journal of Cultural Studies* 25 (1): 86–101.

Watkins, Calvert, ed. 2000. *The American Heritage Dictionary of Indo-European Roots*. New York: Houghton Mifflin Harcourt.

Weerahewa, Jeevika, Gamini Pushpakumara, Pradeepa Silva, Chathuranga Daulagala, Ranjith Punyawardena, Sarath Premalal, Giashudin Miah, Joyasee Roy, Sebak Jana, and Buddhi Marambe. 2012. "Are homegarden ecosystems resilient to climate change? An analysis of the adaptation strategies of homegardeners in Sri Lanka." *APN Science Bulletin* 2: 2–27.

Weiss, Marissa. 2021. "Unexpected threats to trees can be traced to wood pallets." *Feral Atlas*. Accessed June 20, 2022. https://feralatlas.supdigital.org/poster/unexpected-threats-to-trees-can-be-traced-to-wood-pallets.

Wemelsfelder, Françoise, Tony E. A. Hunter, Michael T. Mendl, and Alistair B. Lawrence. 2001. "Assessing the 'whole animal': A free choice profiling approach." *Animal Behaviour* 62 (2): 209–220.

Williams, Raymond. 1976. *Keywords: A Vocabulary of Culture and Society*. London: Fontana Press.

Wisumperuma, Danesh. 2007. "First known record of Guinea grass cultivation in Sri Lanka, 1801–1802." *Journal of the Royal Asiatic Society of Sri Lanka* 53: 219–226.

Wolf, Kathleen L., Sharon T. Lam, Jennifer K. McKeen, Gregory R. A. Richardson, Matilda van den Bosch, and Adrina C. Bardekjian. 2020. "Urban trees and human health: A scoping review." *International Journal of Environmental Research and Public Health* 17 (12): 4371.

Wolfe, Cary. 2017. "Forward." In *Extinction Studies: Stories of Time, Death, and Generations*, edited by Deborah Bird Rose, Thom Van Dooren, and Matthew Chrulew, vii–xvi. New York: Columbia University Press.

Wright, Katherine. 2012. "Rethinking the seasons: New approaches to nature Armidale's imported autumn." *Transformations* 21: 1–14.

Wulf, Andrea. 2015. *The Invention of Nature: The Adventures of Alexander von Humboldt, the Lost Hero of Science: Costa & Royal Society Prize Winner*. London: Hachette UK.

Yates, Frances Amelia. 1966. *Art of Memory*. Abingdon: Routledge.

Yunkaporta, Tyson. 2021. *Sand Talk: How Indigenous Thinking Can Save the World*. New York: HarperOne.

Zucchelli, Christine. 2016. *Sacred Trees of Ireland*. Cork: The Collins Press.

Index

About the Authors

Elizabeth Oriel, PhD, grew up amidst thornless honey locust and crabapple trees in Chicago, Illinois. This is traditional unceded and ancestral homelands of the Potawatomi Nation, and Ojibwe, Odawa, Peoria, Kaskaskia, Miami, Mascouten, Sac and Fox, Kickapoo, Ho-Chunk, and Menomonee tribes lived and/or traded in this area.

Elizabeth works in the environmental humanities as a writer, artist, teacher, and mentor. Her work examines human-wildlife conflict with a more-than-human political lens, as well as human-plant alliances, relations, and futures. Bringing other species' and marginalized voices, interests, and needs to the fore in her writing and artwork is her primary approach. She is currently based at Aarhus University, Denmark.

Anna Perdibon, co-author of chapter 5, is an independent researcher, writer, and storyteller. She explores human-plant relations in ancient, traditional European (Alpine) and Indigenous cultures, and their connections with animism, storytelling, and art. She authored the book *Mountains and Trees, Rivers and Springs: Animist Beliefs and Practices in Ancient Mesopotamian Religion*. She lives in Trentino-Alto Adige, Italy, where she organizes courses, talks, and walks toward an ecological education with plants and their stories.

www.ingramcontent.com/pod-product-compliance
Ingram Content Group UK Ltd.
Pitfield, Milton Keynes, MK11 3LW, UK
UKHW022024310125
454506UK00006B/33